Basic Statistics
for Laboratories

Basic Statistics for Laboratories

A Primer for Laboratory Workers

William D. Kelley
Thomas A. Ratliff, Jr.
Charles Nenadic

JOHN WILEY & SONS, INC.
New York • Chichester • Weinheim • Brisbane • Singapore • Toronto

Published by John Wiley & Sons, Inc., Hoboken, New Jersey.
Originally published as ISBN 0-442-00456-7

For general information on our other products and services please contact our Customer Care Department within the U.S. at 877-762-2974, outside the U.S. at 317-572-3993 or fax 317-572-4002.

Printed in the United States of America
10 9 8 7 6 5 4 3

Library of Congress Cataloging-in-Publication Data

Kelley, William D.
   Basic statistics for laboratories : a primer for laboratory workers / William D. Kelley, Thomas A. Ratliff, Jr., Charles Nenadic.
      p.   cm.
   Includes index.
   ISBN 0-442-00456-7
   1. Statistics.   I. Ratliff, Thomas A.   II. Nenadic, Charles M.
III. Title
QA276.12.K45   1991
519.5—dc20   91-20568
                                                                    CIP

Contents

PREFACE   xi

CHAPTER 1. **Distribution and Variability of Sampling Data**   **1**

        Instructions for Using Random Number Table   4

        FREQUENCY DISTRIBUTIONS   6

        MEASURES OF CENTRAL TENDENCY   9

        MEASURES OF VARIABILITY   11

        THE NEED TO MEASURE VARIABILITY   13

CHAPTER 2. **Measuring Variability**   **13**

        VARIATIONS IN THE NORMAL DISTRIBUTION   17

        IMPORTANCE OF THE NORMAL DISTRIBUTION   20

CHAPTER 3. **Tests of Normality**   **20**

        Example 1: The Poisson Distribution   22

        Example 2: The "Bathtub" Distribution   22

        NORMALITY TESTS   23

CHAPTER 4. Control Charts   28

    INTRODUCTION   28

    SHEWHART CONTROL CHARTS BACKGROUND   28

      Bias   29

      Assumptions   30

    CONTROL CHART PROCEDURES   33

      Initiating the Control Chart   34

      Determine Trial Control Limits   34

      Draw Preliminary Conclusions   34

      Routine Operation   35

    CALCULATIONS   35

    CALCULATION OF CONTROL LIMITS   35

    CALCULATION OF WARNING LIMITS   36

    EXAMPLE—CONTROL CHART FOR REAGENT BLANK
      DETERMINATIONS   37

    References   40

    CONTROL CHARTS FOR INDIVIDUAL VALUES   41

CHAPTER 5. Special Situation Control Charts   41

    MOVING AVERAGES AND RANGES   42

    OTHER CONTROL CHARTS FOR VARIABLES   43

      Variable Subgroup Sizes   44

      R or Sigma (σ) Charts   44

      Cumulative Sum Control Charts   44

      Fraction Defective Charts (p Charts)   47

    INTRODUCTION   49

    TESTS FOR SIGNIFICANCE OF DIFFERENCES   49

CHAPTER 6. **Statistical Tools for the Laboratory**   **49**

References   68

CHAPTER 7. **Sampling Plans**   **69**

References   85

CHAPTER 8. **Correlation and Regression**   **86**

How to evaluate r   92

References   93

CHAPTER 9. **Introduction to Outlying Observations**   **94**

INTRODUCTION   94

BACKGROUND   94

OVERVIEW OF OUTLIER PROBLEMS   95

BASE POPULATION CHARACTERISTICS   98

THE CHOICE OF THE OUTLIER REJECTION TEST   100

OUTLIER TESTS WHEN μ AND σ ARE KNOWN   100

SUMMARY   102

References   102

CHAPTER 10. **Single Outliers**   **104**

INTRODUCTION   104

POPULATION CONTAMINATION   105

WHEN ONLY σ IS KNOWN   105

Example 10.1   106

Example 10.2   108

Example 10.3   110

BOTH μ AND σ UNKNOWN BUT HAVE AN INDEPENDENT
ESTIMATE OF σ^2 111

Example 10.4   111

Example 10.5   117

BOTH μ AND σ UNKNOWN   119

Example 10.6.   120

Example 10.7   120

References   124

CASE 1—ONE HIGH AND ONE LOW OUTLIER   125

CHAPTER 11.   Multiple Outliers   125

TWO HIGH OR TWO LOW OUTLIERS   126

REJECTING MULTIPLE OUTLIERS   129

SKEWNESS AND KURTOSIS TESTS   136

SLIPPAGE TESTS   136

k-Sample Slippage Test (Conover, 1971)   136

YOUDEN EXTREME RANK SUM TEST FOR
OUTLIERS   137

SUMMARY   139

References   141

CHAPTER 12.   Outliers Appearing in Continuous Data Sets   142

INTRODUCTION   142

DUPLICATE MEASUREMENTS   142

ROUTINE TRIPLICATE MEASUREMENTS   143

CONTROL CHARTS   143

USE OF RUNS   144

CONCLUSION   147

REFERENCES   148

CHAPTER 13.   **Statistical Quality Control for Asbestos Counting**   **151**

    INTRODUCTION   151
    AVAILABLE QUALITY CONTROL PROCEDURES   152
    TRADITIONAL TECHNIQUES   152
    REFERENCES   159

CHAPTER 14.   **Nonroutine Quality Control Procedures**   **160**

    SPIKING   161
    REPLICATE ANALYSIS   162
    REFERENCE SAMPLES   163
    BLANK DETERMINATIONS   164
    BLIND SAMPLES   164
    OTHER ANALYTICAL TECHNIQUES   165
    REFERENCES   165

APPENDIX   166

INDEX   171

Preface

This book is designed to provide the laboratory professional worker or technician, who may be uncertain about, or unfamiliar with, the statistical tools used in the control of quality in the laboratory, with the information he or she needs to employ basic statistical controls or reports that may be required by regulatory or accrediting organizations, or may be otherwise useful in the laboratory.

The book tells in simplest terms, step-by-step, how to set up, maintain, and interpret control charts, and explains why they work to the benefit of the user, without the use of complicated mathematics, beyond the high school level. A similar approach is used in explaining other statistical tools commonly used in laboratory work. Simple solutions to other problems involving data handling and interpretation, such as the treatment of outliers, how to deal with single or rarely encountered samples, and so forth, are provided.

Every effort has been made to keep the language of this book nontechnical, and no attempt is made to derive equations or present theory about the various statistical methods explained.

While many of the examples are based on data from industrial hygiene laboratories, the principles involved are applicable to any laboratory testing or analytical problem.

Acknowledgments

We are grateful to the Literary Executor of the late Sir Ronald A. Fisher, F. R. S., to Dr. Frank Yates, F. R. S., and to the Longman Group Ltd, London, for the permission to reprint Tables III and IV from their book *Statistical Tables for Biological, Agricultural and Medical Research* 6th Edition 1974.

Basic Statistics
for Laboratories

1

Distribution and Variability of Sampling Data

PROBABILITIES

Any conclusions reached as a result of test or analytical determinations are charged with uncertainty. No test or analytical method is so perfect, so unaffected by the environment or other external contributing factors, that it will always produce exactly the same test or measurement result or value. Where such situations seem to exist (that is, repetitive tests or analyses giving identical results), either the measurement method is not sensitive enough to detect differences or the person making the measurement is not performing properly.

The sources of variation that contribute to variable results, which will be present in an analytical or test procedure, include: 1) differences among analysts or test technicians, 2) differences in ambient atmospheric conditions, 3) differences between instruments and measurement standards, 4) differences among chemicals, reagents, and test accessories, 5) differences of 1, 2, 3, and 4 over a period of time, and 6) differences in the relationship of 1, 2, 3, and 4 with each other over a period of time. Since this variation must be accepted as a way of life, understanding the nature of variability is extremely important. Fortunately, a "stable system of chance causes" is inherent in the makeup of any process or procedure. This "system of chance causes" will produce a pattern of variation. When this pattern is found to be stable, the process or procedure is said to be "in statistical control," or, more simply, just "in control." Any variation outside this pattern will be found to have an assignable cause that can be determined and corrected. The presence of a "stable system of chance causes" allows for the application of the laws of probability to the statistical treatment and analysis of the data in question.

In discussing probability, we will be dealing with certain statistical "experiments" and their outcomes, which we will call "events." In this frame of reference, any activity or measurement or analytical procedure that generates data will be considered an experiment. Thus, tossing a coin, casting dice, or reading a micrometer or a thermometer are simple forms of experiments.

Consider the simple, repetitive experiment that consists of tossing a coin twice (or its equivalent, tossing two coins simultaneously). In this experiment, there are four, and only four, possible outcomes, as follows:

HH, HT, TH, TT

Or: 2 heads; 1 head, 1 tail; 1 tail, 1 head; and 2 tails.

A simple matrix diagram (Figure 1.1) illustrating this experiment reveals that there is an equal chance of matching or not matching the coins as a result of this experiment.

An event is any particular possible outcome of an experiment. Therefore, the occurrence of two heads (HH) can be considered to be "event A." In the previous experiment, this will occur one time in four, if all of the possible outcomes are equally likely. Another event might be the occurrence of at least one head. Referring Figure 1.1, this can happen three times out of four in the experiment.

Another way of saying this is, "It is probable, in this experiment, that a head will appear three times out of every four times the experiment is repeated."

Probability, then, can be defined as follows: The probability that event A will occur is the ratio of the number of outcomes that involve the appearance of event A to the total number of possible outcomes. Thus, in the experiment of throwing a single, six-sided die, the probability of getting a four is one chance in six or, one-to-six, or expressed another way, 1/6.

The equation for expressing the probability of event A occurring is:

$$P(A) = a/n \qquad (1\text{-}1)$$

Where P(A) is the probability of A occurring
       a is a single outcome
       n is the total number
       of possible outcomes

	H	T
H	HH	HT
T	TH	TT

FIGURE 1.1.   Simple Matrix Diagram.

Applications of probability theory are often concerned with a number of events, rather than a single event. For instance, as a simple example, in the case of two such related events, A_1 and A_2, we may be interested in knowing whether both of these will appear in a particular experiment. This joint event is denoted by the sum $A_1 + A_2$ and its probability by $P(A_1 + A_2)$.

If the two events A1 and A2 possess the property that the occurrence of one prevents the occurrence of the other, they are held to be mutually exclusive events. The results of flipping a coin are examples of mutually exclusive events because the occurrence of a head on any given toss prevents a tail from appearing, and vice versa.

The addition rule which governs the probability of the occurrence of multiple events is expressed as follows:

$$P(A_1 + A_2) = P(A_2) + P(A_2) \text{ when } A_1 \text{ and } A_2 \text{ are mutually exclusive events.} \quad (1\text{-}2)$$

As an example, the probability of getting a two or a four on a roll of a six-sided die is calculated as follows:

$$P(2 + 4) = P(2) + P(4) = 1/6 + 1/6 = 1/3 \quad (1\text{-}3)$$

If A_1 and A_2 are not mutually exclusive events, the probability of $A_1 + A_2$ is given as follows:

$$P(A_1 + A_2) = P(A_1) + P(A_2) - P(A_1 A_2) \quad (1\text{-}4)$$

Where $P(A_1 + A_2)$ is the probability of event A_1 or A_2 occurring
$P(A_1)$ is the probability of event A_1 occurring
$P(A_2)$ is the probability of event A_2 occurring
$P(A_1 A_2)$ is the probability of events A_1 and A_2 occuring

Thus, the probability of getting an ace or a spade on a single draw from a deck of 52 cards is calculated as follows:

$$P(\text{Ace} + \text{Spade}) = 4/52 + 13/52 - 1/52 = 16/52 = 4/13 \quad (1\text{-}5)$$

The reason for having to subtract off P(Ace of Spades) is because we actually counted this event twice when we added P(A) to P(Spade). Note that, in this case, these are not mutually exclusive events.

When conducting an experiment or drawing a sample from a population, care must be taken to preserve complete randomness of the sample, insofar as possible. All bias must be rigorously avoided when selecting samples. An example of a bias would be, when flipping a coin, to always start with the head up. Other examples are selecting documents for audit from only one drawer of file cabinet or drawing test samples from only the top layer of one packing box out of a lot consisting of several packing boxes.

Selecting random samples can be facilitated by using a table of random numbers (Table 1.1) or by using random numbers obtained from a random number generator computer

Instructions for Using Random Number Table

A typical scenario for using a random number table follows.

You wish to take a sample of calibration record documents from a file cabinet that contains 500 documents in each drawer. Referring to Tables I & IIA pages 78 & 79 you determine that the sample size for a population of 2000 is 126. This means that you should examine every sixteenth document. To start the random selection process, enter the random number table using a pencil, with your eyes closed. If your pencil lands on the digit "9," go to the ninth row down in the table and count over nine digits. Find the digit "5." In the first drawer (not necessarily the top drawer), start with the 5th document from the front and then take every 16th document as a sample. Repeat this procedure for every drawer. You may elaborate on this scheme, to further randomize the sample selection plan.

Randomizing sample selection is not always easy. Consider the problem of selecting a sample of 1250 tablets from a production lot of 500,000 tablets packaged in 50 tablet bottles and packed in cartons of 24 bottles each. This sample selection problem requires expensive, time-consuming effort. Less sophisticated randomness may be obtained by the use of a pair of dice, flipping a coin, and so on.

In the experiments discussed up to this point, using coins, dice, and playing cards, solid information has been available about the nature of the experiment, that is, we know that there are two sides to a coin, six sides to the die being used, and 52 cards in a deck. However, in real life in the laboratory, such information about the nature of samples is usually not available. In addition, as seen, the samples and the methods used to analyze or test them are subjected to many variables. Because so much information is lacking, probabilities must be obtained empirically and predictions made based on past performance. It is important to remember that such predictions are approximate and are made only with a certain degree of confidence in their accuracy. (See Chapter 4 for a discussion of the terms "accuracy, precision, and bias.")

What statistical techniques do offer is more information about the data in question, so that an informed decision can be made based on all available knowledge. These variables make it necessary to use empirical probabilities in decision making, which may often lead to error. The term "error" in this context is obtaining a result that is not the true value. While the laboratory practitioner may suspect that

TABLE 1.1.

	00-04	05-09	10-14	15-19	20-24	25-29	30-34	35-39	40-44	45-49
00	54463	22662	65905	70639	79365	67382	29085	69831	47058	08186
01	15389	85205	18850	39226	42249	90669	96325	23248	60933	26927
02	85941	40756	82414	02015	13858	78030	16269	65978	01385	15345
03	61149	69440	11286	88218	58925	03638	52862	62733	33451	77455
04	05219	81619	10651	67079	92511	59888	84502	72095	83463	75577
05	41417	98326	87719	92294	46614	50948	64896	20002	97365	30976
06	28357	94070	20652	35774	16249	75019	21145	05217	47286	76305
07	17783	00015	10806	83091	91530	36466	39981	62481	49177	75779
08	40950	84820	29881	85966	62800	70326	84740	62660	77379	90279
09	82995	64157	66164	41180	10089	41757	78258	96488	88629	37231
10	96754	17676	55659	44105	47361	34833	86679	23930	53249	27083
11	34357	8804	053364	71726	45690	66334	60332	22554	90600	71113
12	06318	37403	49927	57715	50423	67372	63116	48888	21505	80182
13	62111	52820	07243	79931	89292	84767	85693	73947	22278	11551
14	47534	09243	67879	00544	23410	12740	02540	54440	32949	13491
15	98614	75993	84460	62846	59844	14922	48730	73443	48167	34770
16	24856	03648	44898	09351	98795	18644	39765	71058	90368	44104
17	96887	12479	80621	66223	86085	78285	02432	53342	42846	94771
18	90801	21472	42815	77408	37390	76766	52615	32141	30268	18106
19	55165	77312	83666	36027	28420	70219	81369	41943	47366	41067
20	75884	12952	84318	95108	72305	46420	91318	89872	45375	85436
21	16777	37116	58550	42958	21460	43910	01175	87894	81378	10620
22	46230	43877	80207	88877	89380	32992	91380	03164	98656	59337
23	42902	66892	46134	01432	94710	23474	20423	60137	60609	13119
24	81007	00333	39693	28039	10154	95425	39220	19774	31782	49037
25	68089	01122	51111	72373	06902	74373	96199	97017	41273	21546
26	20411	67081	89950	16944	93054	87687	96693	87236	77054	33848
27	58212	13160	06468	15718	82627	76999	05999	58680	96739	63700
28	70577	42866	24969	61210	76046	67699	42054	12696	93758	03283
29	94522	74358	71659	62038	79643	79169	44741	05437	39038	13163
30	42626	86819	85651	88678	17401	03252	99547	32404	17918	62880
31	16051	33763	57194	16752	54450	19031	58580	47629	54132	60631
32	08244	27647	33851	44705	94211	46716	11738	55784	95374	72655
33	59497	04392	09419	89964	51211	04894	72882	17805	21896	83864
34	97155	13428	40293	09985	58434	01412	69124	82171	59058	82859
35	98409	66162	95763	47420	20792	61527	20441	39435	11859	41567
36	45476	84882	65109	96597	25930	66790	65706	61203	53634	22557
37	89300	69700	50741	30329	11658	23166	05400	66669	48708	03887
38	50051	95137	91631	66315	91428	12275	24816	68091	71710	33258
39	31753	85178	31310	89642	98364	02306	24617	09609	83942	22716
40	79152	53829	77250	20190	56535	18760	69942	77448	33278	48805
41	44560	38750	83635	56540	64900	42912	13953	79149	18710	68618
42	68328	83378	63369	71381	39564	05615	42451	64559	97501	65747
43	46939	38689	58625	08342	30459	85863	20781	09284	26333	91777
44	83544	86141	15707	96256	23068	13782	08467	89469	93842	55349
45	91621	00881	04900	54224	46177	55309	17852	27491	89415	23466
46	91896	67126	04151	03795	59077	11848	12630	98375	52068	60142
47	55751	62515	21108	80830	02263	29303	37204	96926	30506	09808
48	85156	87689	95493	88842	00664	55017	55539	17771	69448	87530
49	07521	56898	12236	60277	39102	62315	12239	07105	18844	01117

Reprinted by permission from Statistical Methods, Eighth Edition, by G. Snedecor and W. Cochran © 1989 by Iowa State University Press, Ames. Iowa.

a result is in error, more often than not there will be no means by which the existence or extent of the error can be confirmed.

Errors are classed as determinate (having assignable causes) and indeterminate (random, without apparent cause) Determinate errors are:

Method errors
Personal errors
Instrument errors

It is useful to identify determinate errors because corrective action can be taken to remove them as sources of unwanted incorrect results.

Determinate errors lead to laboratory bias, since they constantly affect laboratory results in one direction. They also destroy the necessary "stable system of chance causes" that allows the laboratory to safely use the laws of probability.

This is one of the reasons that we are concerned with understanding how the laws of probability operate. The other is that our dealings with statistical methods will be focussed on using control charts. The usefulness of control charts is predicated on the assumption that the data to be monitored or "controlled" will be normally distributed. The determination as to the normality of data is based on the use of the principles of probability theory. Before investigating the nature of the normal distribution, a discussion of distribution of data is in order.

FREQUENCY DISTRIBUTIONS

Frequency distributions present, in tabular form, the number of times a given value occurs in a data set. When a naturally occurring phenomenon is measured, as is often the case in laboratory work, the results, if tabulated, will tend to form what is called a normal, or Gaussian, distribution. As an example, Figure 1.2 is a table showing the results of a simple experiment yielding 100 data points, with values from 1 to 11. This kind of table, which is a form of frequency distribution, is called a histogram. For the purposes of this illustration, the data points were selected to show a perfectly normal distribution, around the point of central tendency, the area in the distribution pattern where most of the observations seem to accumulate. Deviations from this pattern will be discussed later in this chapter. If straight lines are drawn between the right-hand ends of the histogram columns, the resulting figure is called a distribution polygon (Figure 1.3), a and if a smooth curve is formed by joining the ends of the column, the resulting figure is called a distribution curve (Figure 1.4). In this case, the curve is for a normal or Gaussian distribution.

Variations in the shape of the normal curve are often encountered, depending on how the data points are concentrated around the center of, and toward the tails of, the curve. There are four characteristics of a set of data that are illustrated by a

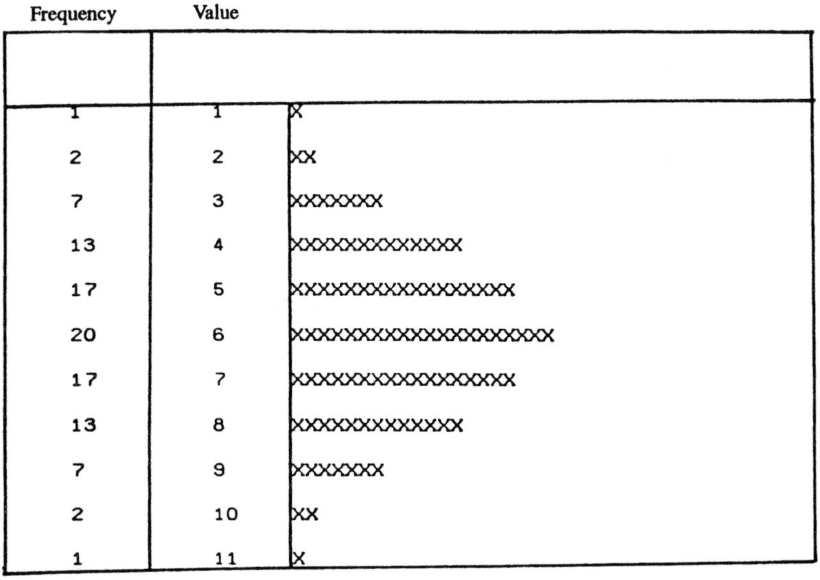

Frequency	Value	
1	1	X
2	2	XX
7	3	XXXXXXX
13	4	XXXXXXXXXXXXX
17	5	XXXXXXXXXXXXXXXXX
20	6	XXXXXXXXXXXXXXXXXXXX
17	7	XXXXXXXXXXXXXXXXX
13	8	XXXXXXXXXXXXX
7	9	XXXXXXX
2	10	XX
1	11	X

FIGURE 1.2.   Distribution Histogram.

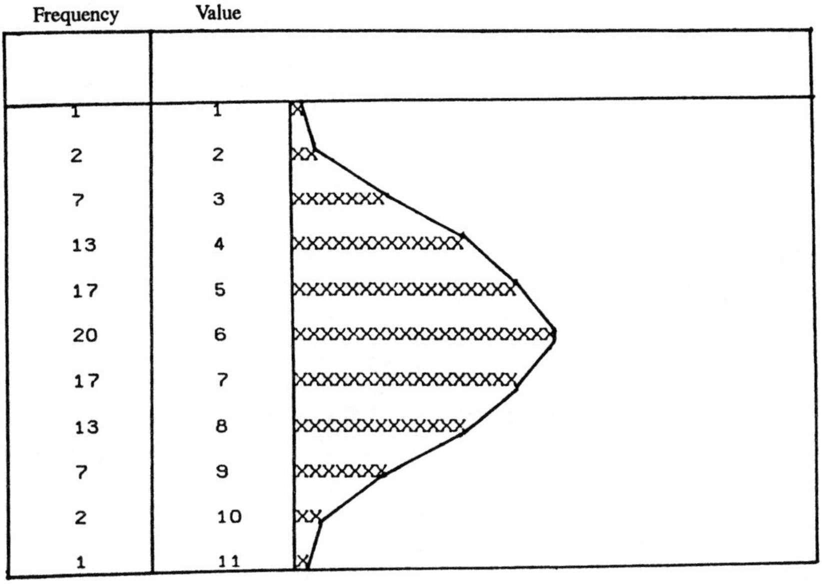

Frequency	Value	
1	1	X
2	2	XX
7	3	XXXXXXX
13	4	XXXXXXXXXXXXX
17	5	XXXXXXXXXXXXXXXXX
20	6	XXXXXXXXXXXXXXXXXXXX
17	7	XXXXXXXXXXXXXXXXX
13	8	XXXXXXXXXXXXX
7	9	XXXXXXX
2	10	XX
1	11	X

FIGURE 1.3.   Distribution Polygon.

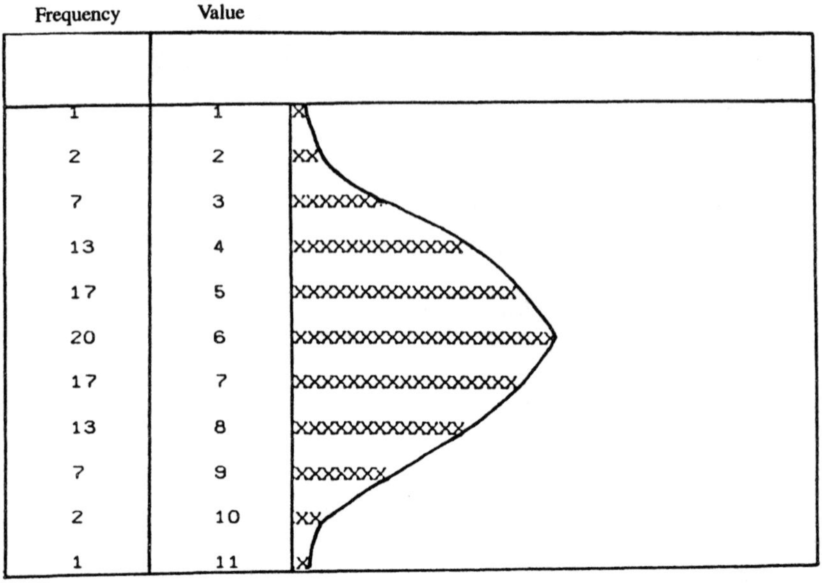

Frequency   Value

Frequency	Value
1	1
2	2
7	3
13	4
17	5
20	6
17	7
13	8
7	9
2	10
1	11

FIGURE 1.4.   Distribution Curve.

distribution curve. The first is how the data values cluster around the highest point, or points, in the curve. Some data sets have more than one focal point of concentration. In such cases, they are called "multimodal distributions." If the distribution histogram or curve has only one high point, it is "unimodal." The characteristic describing where the data points fit around the center of the distribution is called a "measure of central tendency."

The second characteristic one may observe is how widely the data points are scattered around the center of the distribution. This characteristic is called a measure of variability." If the distribution curve is flattened out and extends widely to each side, the data demonstrates a greater pattern of variation than one whose curve has a sharp peak and a narrow base.

The third characteristic demonstrated by a distribution curve is the symmetry of the the curve about the center — that is, whether the data points are evenly arranged to the left and right of the measure of central tendency. If they are not, the resulting curve is said to be "skewed." If the tail of the distribution curve runs farther to the right than to the left, the curve is "positively skewed." If extended in a similar manner to the left, it is negatively skewed."

The last characteristic of distribution curve to be considered is its "kurtosis." Kurtosis is the measure of flatness or peakedness of the distribution curve.

For the purposes of this text, it suffices to point out that the laboratory practi-

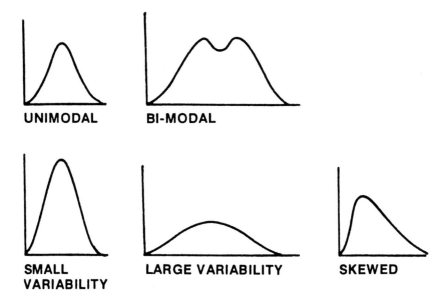

UNIMODAL   **BI-MODAL**

**SMALL   LARGE VARIABILITY   SKEWED
VARIABILITY**

FIGURE 1.5.   Variations in the Distribution Curve.

tioner may encounter variations in the appearance of the normal curve. Evaluating and appraising the degree of non-normality will be covered in a later chapter. For examples of variations in the shape of the normal curve, see Figure 1.5.

MEASURES OF CENTRAL TENDENCY

There are three principal measures of central tendency: the mean, the mode, and the median. Of these, the most commonly used is the mean or arithmetic average. The mean is simply the total of all the values in a data set divided by the number of values. The mean is the most useful of all the measures of central tendency because it uses all the values in a set of results. The notation for the sample[1] mean is $\overline{X}$. Similarly, for the population[2] mean, of which $\overline{X}$ is an estimate, the notation is μ. The equation for calculating the mean is:

$$\text{Mean---} \quad \overline{X} = \frac{\Sigma X}{n} \tag{1-6}$$

---

[1]Sample—A sample is a lesser quantity of objects or measurements drawn from a larger population (or universe), preferably in a random manner.

[2]Population—The larger group (or total group) from which a sample is drawn.

As an example, the mean is calculated as in Table 1.2.

Another common measure of central tendency is the median (M_1. The median is the point in a data set where there are as many values occurring above it as there are below it. To find the median of the data in Table 1.2, we need only to arrange the figures in order, from the lowest to the highest (or vice versa), and pick the middle value. When there is an even number of observations, the average of the two middle values becomes the median value. For the data in Table 1.2, the median is 56.

The mode, M, is another measure of central tendency, although it is of little practical importance. The mode is the most frequently occurring value in a data set; therefore, it is the value at the peak of the frequency distribution curve, and is very easy to determine. Some data will have more than one mode. This distribution is called bimodal, in the case of two modes, or multimodal, when more than two modes occur. In the data in the Table 1.2, no two data points repeat themselves, so,

TABLE 1.2.   Calculating the Mean.

Test No.	Results in Ft./Lbs.
1	53
2	72
3	59
4	45
5	44
6	85
7	77
8	56
9	157
10	83
11	120
12	81
13	35
14	63
15	48
16	180
17	94
18	110
19	51
20	47
21	55
22	43
23	28
24	38
25	26
Total	1750

$$\overline{X} = \frac{\Sigma X}{n} = \frac{1750}{25} = 70.0 \qquad (1\text{-}7)$$

in order to estimate the mode, we can group the data. If we form a frequency table for the data above, the group 40–55 forms a high peak. If we take the midpoint of this interval, 47.5, we have a crude mode for the data. In a symmetrical curve, such as Figure 1.4, the mode and the median are equal.

MEASURES OF VARIABILITY

As with measures of central tendency, there are several measures of variability or dispersion, the most common of which is the standard deviation. The population standard deviation is designated by the Greek letter σ, and the sample standard deviation, which provides an estimate of σ, is denoted by the lower case "s."

The equation for the population standard deviation is:

$$\sigma = \sqrt{\frac{\Sigma(X_1 - X)^2}{n}} \qquad (1\text{-}8)$$

The equation for the sample standard deviation is:

$$s = \sqrt{\frac{\Sigma(X_1 - X)^2}{n-1}} \qquad (1\text{-}9)$$

The standard deviation for the sample data above is computed as follows:

$$\Sigma\, s^2 = (53 - 70)^2 + (72 - 70)^2 + (59 - 70)^2 + \qquad (1\text{-}10)$$
$$(38 - 70)^2 + (26 - 70)^2 = 35{,}146$$

$$\frac{\Sigma\, s^2}{n-1} = \frac{35{,}146}{24} = 1464.4$$

$$s = \sqrt{(1464.4)} = 38.28$$

The sample value s^2 is referred to as the "variance." The variance is the square of the standard deviation and is another measure of variability. The variance of the population is known as "σ^2". The variance is useful when it is necessary to determine the average standard deviation. Variances can be summed and divided by their number, n, taking the square root of the result to obtain the average standard deviation. Standard deviations cannot be added and the sum divided in the same manner. Variances result from and measure the relative components of variability.

The range, R, is also used as a measure of variability. It is simply the lowest value subtracted from the highest, in a set of data. For the data in Table 1.2, the lowest value is 26 and the highest is 180. The range is then:

$$R = \text{Max} - \text{Min} = 180 - 26 = 154$$

The reason the range is of interest as a measure of variability is that, for small sets of data, it is much easier to calculate than the standard deviation, while still offering a reasonably precise estimate of sigma. The method for doing this will be explained in Chapter 6.

References

Enrick, N. L., and H. E. Motley. 1966. *Manufacturing Quality Control.* Newark, NJ: General Instrument Corp.

Kelley, W. D. 1968. Analytical quality control—Radiological. In *Statistical Method—Evaluation and Quality Control for the Laboratory.* Cincinnati: Public Health Service.

Lambert, M. L. 1968. Probability theory and statistical distributions. In *Statistical Method—Evaluation and Quality Control for the Laboratory.* Cincinnati: Public Health Service.

Lambert, M. L. 1968. Measures of central tendency and dispersion. In *Statistical Method—Evaluation and Quality Control for the Laboratory.* Cincinnati: Public Health Service.

Snedecor, George W. 1956. *Statistical Methods.* Ames, IA: The Iowa State College Press.

——July 1983. *Industrial Hygiene Laboratory Quality Control.* Cincinnati: National Institute for Occupational Safety and Health.

2

Measuring Variability

THE NEED TO MEASURE VARIABILITY

Knowing the mean value of a set of data does not really give enough information about it. The effect of this lack of information is pointed out in the following example.

Two chemists analyze aliquots of the same material, for which the correct value is estimated to be approximately 20 to 25 units. They both report results having a mean value of 23.5 units. With this limited information, there would be no way to pick which of the chemists did the better job of analyzing the samples, because they both arrived at exactly the same value for their sets of data and both were close to the estimate of the correct value.

Suppose, however, that each chemist had analyzed the four samples to arrive at the same mean value, but had gotten results as follows:

Chemist A: 20, 23, 25, and 26 units.
Chemist B: 10, 40, 15, and 29 units.

Adding the four sample results for each of these, and dividing by four, we arrive at a sample mean of 23.5 for each, yet it is obvious that Chemist A is doing a better job. Despite the fact that both chemists arrived at the same mean value, looking at the individual measurements shows that there is a great deal of difference in the confidence that one can place in the precision of the two different set of results. It is obvious that the precision of Chemist B leaves something to be desired.

The results of Chemist A show a spread of only 6 units, while that of Chemist B has a spread of 30 units. From this example, we can conclude that having

13

information about an average result is not sufficient and that we need to know more about the nature of the data set.

The last chapter spoke about the importance of the standard deviation. In order to calculate the standard deviation of a sample we will use the equation:

$$s = \sqrt{\frac{\Sigma(X_i - \overline{X})^2}{n-1}} \qquad (2\text{-}1)$$

As a simple illustration, we will use the following data and substitute values in the equation:

X	X - $\overline{X}$	(X - $\overline{X}$)^2
1	-2	4
2	-1	1
3	0	0
4	1	1
5	2	4
$\Sigma X = 15$		$\Sigma(X - \overline{X})^2 = 10$

$$\sigma = \sqrt{\frac{10}{5}} = \sqrt{2} = 1.414$$

$$\overline{X} = \frac{\Sigma X}{n} = \frac{15}{5} = 3 \qquad (2\text{-}2)$$

Transposing the terms in the equation for the sample standard deviation as follows, gives us the ability to simplify the calculations, while arriving at the same result. Examples of the two methods of performing these calculations appear below:

X	X - $\overline{X}$	(X - $\overline{X}$)^2
1	-2	4
2	-1	1
3	0	0
4	1	1
5	2	4
15		10

$\Sigma X = 15$

$n = 5$

$\overline{X} = 3$

$(X - \overline{X})^2 = 10$

$$s = \sqrt{\frac{\Sigma(X_1 - \overline{X})^2}{n-1}} = \sqrt{\frac{10}{4}} = \sqrt{2.5} = 1.6 \qquad (2\text{-}3)$$

X	X^2
1	1
2	4
3	9
4	16
5	25
15	10

$\Sigma X = 15$

$n = 5$

$\Sigma X^2 = 55$

$$s = \sqrt{\dfrac{\Sigma X^2 - \dfrac{(\Sigma X)^2}{n}}{n-1}} = \sqrt{\dfrac{55 - \dfrac{(15)^2}{5}}{4}} = \sqrt{\dfrac{55 - \dfrac{225}{5}}{4}} = \qquad (2\text{-}4)$$

$$\sqrt{\dfrac{55-45}{4}} = \sqrt{\dfrac{10}{4}} = \sqrt{2.5} = 1.6$$

As said in the first chapter, in order for the sampling data to be meaningful, the samples must be drawn at random. That is, each sample is drawn from the population in such a way that all samples have an equal chance of being picked from the population or lot. The sampling data is then used to estimate the nature of the population (Figure 2.1). Note that the term "n - 1" is used in the denominator in equation 2-1, instead of the term "n." In doing this, one degree of freedom is removed. Here we will define a "degree of freedom" as a constraint upon the system. The number of degrees of freedom used when computing the sample standard deviation is usually the number of samples taken minus one. The one degree of freedom (the constraint on the system) subtracted here is the one constant, the mean, which must be calculated in order to arrive at the standard deviation.

There is not total agreement among authorities about the accepted notation for the population standard deviation and the sample standard deviation. Although σ and s are the most widely used symbols, the reader may run into variations from time to time, and should not be confused by them.

As seen in Chapter 1, the variance is computed by squaring the standard deviation (i.e., Variance = σ^2 A related measure of variability is "relative standard deviation" (RSD), or the "coefficient of variation" (CV), which is the standard deviation divided by the mean:

$$CV = \dfrac{\sigma}{X} \qquad (2\text{-}5)$$

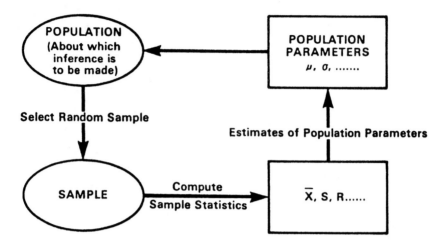

FIGURE 2.1.   Relationship Between Sample Statistics and Population Parameter.

To express the fraction as a percentage, multiply the result by 100. The coefficient of variation is often used as an expression of variability, when several sets of results are similar in their degree of variability.

For the illustration in 2-2, the variance, then, is $\sigma^2 = 1.414^2 = 2$, and the coefficient of variation is

$$CV = 1.414/3 = .471(100) = 47\%. \qquad (2\text{-}6)$$

Another technique for simplifying the calculation of the standard deviation is to code the data as shown in Table 2.1. By removing the decimal points and subtracting 750 from each data entry as shown, only small numbers are left to handle. By doing this, we compute a coded σ of five units; then, by replacing the decimal point in its proper position, we obtain a true σ of .005.

A note of caution regarding the use of programmable calculators when computing the standard deviation: Programs can give either population or sample standard deviations, so it is important to know which result your calculator furnishes when dealing with small sample sizes. The examples shown in Figures 2.1 and 2.2 can be used as an easy check on how the input to one's calculator is treated. In order to verify which standard deviation is programmed, run the data shown in the examples. If the result is 1.414, it has computed a population standard deviation; if the result is 1.6, it is the sample standard deviation.

TABLE 2.1.   Calculation of the Standard Deviation Using Coded Data.

X	X (Coded)	$X - \overline{X}$	$(X - \overline{X})^2$
.750	750	0	0
.746	746	-4	16
.744	744	-6	36
.753	753	3	9
.749	749	-1	1
.759	759	0	0
.747	747	-3	9
.745	745	-5	25
.762	762	12	144
.754	754	4	16
.748	748	-2	4
.754	754	4	16
	9002		276

$$\Sigma X = 9002$$
$$n = 12$$
$$\overline{X} = 750.1 = 750$$
$$\Sigma(X - \overline{X})^2 = 276$$

$$s = \sqrt{\frac{\Sigma(X - \overline{X})^2}{n-1}} = \sqrt{\frac{276}{11}} = \sqrt{25.09} = \sqrt{25} = 5 \text{ (coded)} \qquad (2\text{-}7)$$

Note: A word of caution about the calculation of σ. Do not round off until computations are completed. Rounding off too soon may result in the loss of significant figures when the square root is taken.

VARIATIONS IN THE NORMAL DISTRIBUTION

Normal (Gaussian) distributions may vary in form or shape. Looking at Figure 2.2, we see three distribution curves, all having the same mean. Note that the spread and height of the three distributions are dependent on the difference in the standard deviations of the data being, from top to bottom, 0.25, 0.5, and 2.0, respectively.

This illustrates how the standard deviation acts as a measure of variability—the larger the spread of the distribution curve, the larger the standard deviation. If the curve is narrower and more pointed, the standard deviation is smaller. All three of these curves have exactly the same mean.

Looking at Figure 2.3, we find two distributions that have the same standard deviation and two different means. In this case, the shape of the curves is the same, but they appear in separate locations along the horizontal axis.

When both the standard deviation and the mean are changed, we get the effects shown in Figure 2.4. In the top set of curves, we have two different means and two different standard deviations. If we reverse the situation, we get the effect shown in the bottom set of curves.

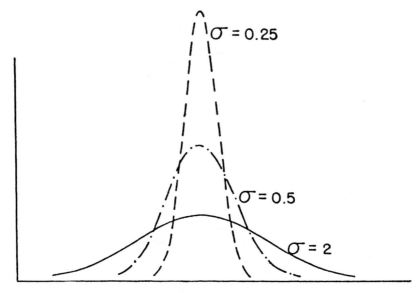

FIGURE 2.2.   Effect of Changing σ Alone and Keeping μ the Same.
σIs as Measure of Spread or Variation in Original Scale Units.

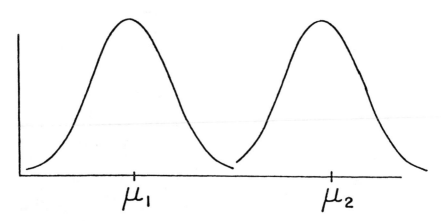

FIGURE 2.3.   Effect of a change in μ.
μ₂ Larger than μ₁.
σ Same for the Curves.

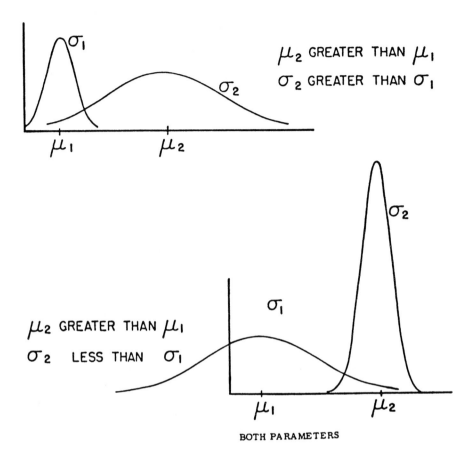

FIGURE 2.4.   Effect of changing both μ & σ.

References

Duncan, A. J. 1965. *Quality Control and Industrial Statistics*. Homewood, IL: Richard D. Irwin, Inc.

Schrock, E. M. 1947. *Quality Control and Statistical Methods*. New York: Reinhold Publishing Corporation.

3

Tests of Normality

IMPORTANCE OF THE NORMAL DISTRIBUTION

In the previous chapters, we have discussed the nature of the normal distribution in some detail. The reason for the emphasis on this subject is that the usefulness of $\overline{X}$-R control charts is based on the assumption that the sample averages of the underlying data plotted on the charts will tend to be normally distributed. Another reason the normal distribution is important is that many other distributions can be approximated by the normal distribution. The theory supporting this concept is known as the "Central Limit Theorem." Stated in its simplest terms, this theorem states that:

Regardless of the shape of the distribution of the population from which the samples are drawn, the distribution of the averages, X, of the subgroups or samples will tend to be normally distributed. Further, in a normal distribution, 99.9% of the data points will fall within plus-or-minus three standard deviations from the mean (Figure 3.1). Thus, the probability of an occurrence falling outside the three sigma limit is very remote. On each side of X, the mean, the data points will fall as follows:

± 1 standard deviation—68.27%
± 2 standard deviations—95.44
± 3 standard deviations—99.745%

This can be used to predict what percentage of data points might fall within or outside of the distribution curve. Before discussing how to determine if a data set is normally distributed, it would be beneficial to point out that there are many other kinds of distribution patterns, whose shapes may be vastly different from that of the normal or Gaussian curve. To list just a few of these distributions, there are:

20

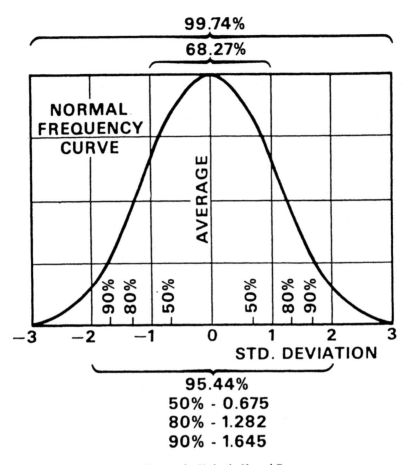

FIGURE 3.1.   Percentage of Frequencies Under the Normal Curve.

Binomial	Poisson	F
Hypergeometric	t	Chi Squareκ^2
Weibull	"Bathtub"	Lognormal.

distributions among others. Of these, the lognormal distribution is often found in environmental air measurements. (The skewed curve in Figure 1.5 represents a lognormal distribution.) To illustrate how other distribution curves may differ radically from the normal, we will show just two of these, so that the reader may become acquainted with them, since they may be encountered from time to time in laboratory work. A detailed explanation of the use and derivation of the equations describing these distributions is beyond the scope of this text.

Example 1: The Poisson Distribution

This data distribution is named for the French mathematician, Simeon Poisson (1781–1840), known for his work on probability. The underlying theory governing the use of the Poisson distribution usually involves the occurrence of small numbers of events out of a great number of possibilities for that event to happen. The Poisson distribution is useful in the laboratory for particle counting and radiation work. For our example, we will construct an imaginary scenario.

In order to plan for the needs for replacement horseshoes in the coming year, the regimental veterinary of a horse cavalry regiment records the loss of horseshoes during a day-long road march of the 5th Cavalry Regiment, in which 1000 troopers participate. During the march, the horses lose shoes as shown in the following table:

No. of Shoes Lost	No. of Horses Losing Shoes
0	802
1	108
2	52
3	31
4	7
	1000

This results in the curve shown in Figure 3.2.

Example 2: The "Bathtub" Distribution

The "bathtub" distribution is associated with product or instrument reliability and is a pictorial representation of instrument failure rate. Laboratories using remote recording instruments are, of course, extremely interested in the performance (i.e., the failure rate) of such unattended equipment. The curve for the bathtub distribution resembles the cross-section of a bathtub if one were to saw it across its middle. A typical bathtub curve is characterized by an initial period of high failure rates, during which time the problems causing failures are debugged. This corrective action causes the failure rate to diminish and phase into a constant or stable rate of failure that continues until the end of the instrument or product life is approached. At this point, the "wear-out" period begins and the failure rate rises again. This results in the curve shown in Figure 3.3.

Again, these figures are provided merely to illustrate the differences in distribution curves that may be encountered.

However, we are primarily concerned with the Gaussian or normal distribution, since its characteristics are basic to the construction, use and interpretation of control charts, as will be seen in the following chapters.

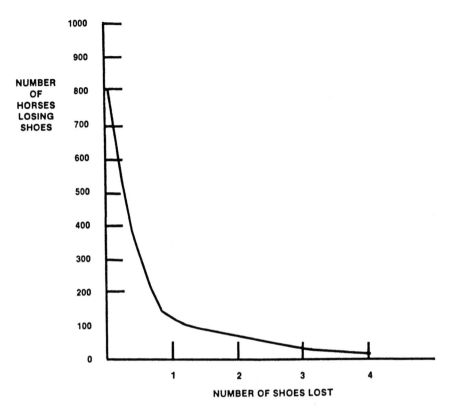

FIGURE 3.2.   Horseshoe Loss During Road March.

NORMALITY TESTS

It is often wise, if possible, to test this assumption of normality. Various procedures are available for making this test. For the examples shown here, we will use the following table, which gives the frequency distribution of the results of a series of 145 similar tests (Table 3.1).

If there are a sufficient number of samples, the simplest method of determining the normality of the distribution of a data set is to construct a histogram. The next step is to compare the histogram with a normal curve that has the same mean and standard deviation, to see how well they compare. Such a histogram, using the data from Table 3.1, is shown in Figure 3.4.

This is an imprecise method at best, but has the advantage of being simple and easy to accomplish. However, unless there is an extremely good fit of a normal curve laid over the distribution polygon, it is better to plot the cumulative percent-

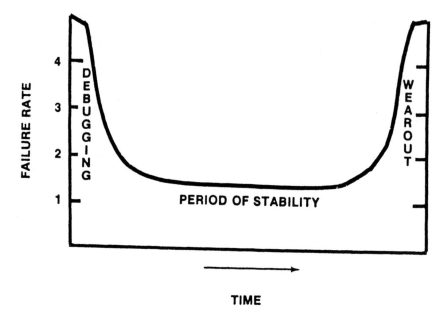

TIME

FIGURE 3.3.   The Bathtub Curve.

TABLE 3.1.   Table of Test Results

Grams	Observed Frequency	Percentage	Cumulative
.8485	2	1.38	1.38
.8455	1	.69	2.07
.8425	2	1.38	3.45
.8395	6	4.14	7.59
.8365	7	4.83	12.42
.8335	23	15.86	28.28
.8305	55	37.93	66.21
.8275	21	14.48	80.69
.8245	14	9.66	90.35
.8215	5	3.45	93.80
.8185	4	2.76	96.56
.8155	3	2.07	98.69
.8125	2	1.38	100.01
TOTAL	145	100.01	100.01

*Adapted from Table 4, QUALITY CONTROL AND INDUSTRIAL STATISTICS A. J. Duncan, Rev. Ed. Used with permission of Richard D. Irwin, Inc.

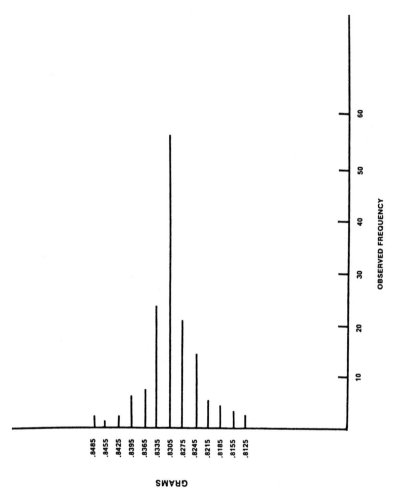

FIGURE 3.4.

25

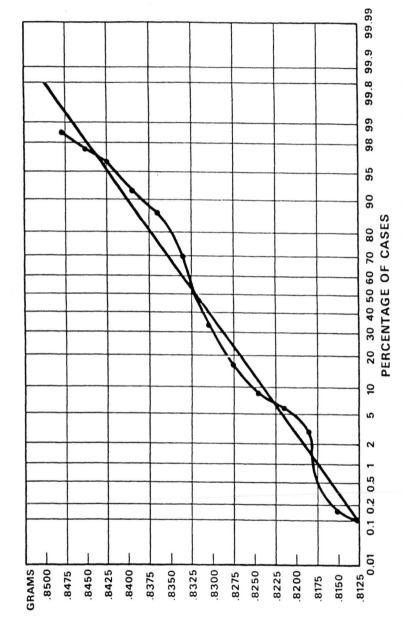

FIGURE 3.5. Adapted from Figure 133, *Quality Control and Industrial Statistics*, by A. J. Duncan. Rev. Ed. Used with the permission of Richard D. Irwin, Inc.

26

age points of the distribution on normal probability paper. Normal probability paper is numbered from 0.01% to 99.99% on the abscissa, with the ordinate scale left blank for entering the data points being plotted, To use the probability paper, the first step is to enter the data points on the vertical axis, using appropriate scale intervals. Next, compute the percentage of each of the observed frequencies from the data. After obtaining the percentages, as shown in the third column of Table 3.1, the Cumulative Percentage Points are calculated by successively adding each of the percentages of the observed frequencies, as shown in the fourth (right-hand) column of Table 3.1. Having obtained the Cumulative Percentage Points, the next step is to plot them on the horizontal axis. After the plot is completed, a curve is drawn through the data points. The last step is to draw a best-fit line through the data points, using a straight edge. The result of such a plot is shown in Figure 3.5.

The curve of this plot of the cumulative percentage points shows some departure from the best-fit straight line. Whether this degree of misfit is significant cannot be told from merely looking at this graph. However, a Chi-square (χ^2) test for goodness of fit tells us that the distribution of this data is not normal. This example was selected so that it can be used as a point of departure for judging any plot that the reader may develop from his or her own data. In making this judgment, one must assume that the closer the plotted curve follows the straight best-fit line, the nearer the data distribution is to normality.

When making the best-fit estimate, as a rule of thumb, the user should ignore data points below 10% and above 90%, as it will be found that anomalies often occur in these areas of the curve.

It should be pointed out here that the most accurate way of testing for normality is to use the χ^2 test for normality of data. However, the calculations are tedious and time consuming for desk calculator computation. Standard χ^2 computer programs are available. However, judgment must be used in weighing the cost of getting an accurate determination against the value of the information.

References
Duncan, A. J. 1955. *Quality Control and Industrial Statistics.* Homewood, IL: Richard D. Irwin, Inc.

Enrick, N. L. 1972. *Quality Control and Reliability.* New York: Industrial Press, Inc.

4

Control Charts

INTRODUCTION

One of the most generally applicable and easily applied statistical quality control techniques is the Shewhart Control Chart. The Shewhart Control Chart can be applied to almost any area of testing, analysis, calibration, or research.

The application of control chart techniques to routine testing or analysis data requires some adaptation and translation from the industrial manufacturing practices for which the control chart was originally developed. Control charts were first developed to control production processes, where large numbers of articles were being manufactured and inspected on a continuous basis. Testing and analytical work produce fewer results on an intermittent basis. Therefore, the test engineer, analyst, or technician, having less data to work with, must make certain concessions in order to respond quickly to undesired changes in procedures.

In all cases, though, an understanding of the basic principles of using Shewhart Control Charts will enable the test technician or analyst to efficiently maintain the precision and accuracy of the results.

SHEWHART CONTROL CHARTS
BACKGROUND

The Shewhart Control Chart was devised by Dr. Walter Shewhart of the Bell Telephone Laboratories. The description of the design, use, and interpretation of control charts was published in 1931 in Shewhart's book, *Economic Control of Quality of Manufactured Product*. This statistical technique did not at first gain wide acceptance. However, at the present time, it is perhaps the most useful and widely used statistical tool in the analytical or testing laboratory.

The Shewhart Control Chart (commonly abbreviated "control chart") can serve several functions. The control chart can be used as a tool to:

1. Determine and define empirically acceptable levels of quality of laboratory performance.
2. Achieve the acceptable level defined.
3. Maintain performance at that acceptable level of quality.

Control charts can be used to control:

1. Measurable quality characteristics.
2. The fraction defective or percent defective in a sample.
3. The number of defects found per unit.

This discussion, however, will deal primarily with control charts for measurable quality characteristics.

The control chart is a tool for distinguishing the pattern of indeterminate (random) error or variation from the determinate (assignable cause) error or variation. This technique displays the test data from a process or method in a form that graphically compares the variability of all test or analytical results with the expected or average variability of small groups of data, giving, in effect, a graphical analysis of variance that is a comparison of "within groups" variability versus the "between group" variability.

To illustrate the difference between determinate and indeterminate errors, the following example is given. Numbers can be employed to either enumerate objects or describe quantities. For instance, if 16 air samples are taken simultaneously in different locations in a warehouse where gasoline powered forklifts are moving about, the number or count of lift trucks would be the same regardless of who counted them, when the count was made, or how the count was made. However, when each individual sample of air is analyzed for carbon monoxide, 16 different results or numbers would be obtained.

As seen in Chapter 1, experimental errors are classified as determinate or indeterminate. A count of samples yielding a result of 15 would be a determinate error, which would be quickly disclosed by a recount. An indeterminate error, however, would result from the inherent variability in repetitive determinations for carbon monoxide by gas chromotography, infrared, or a colorimetric method.

Bias

Bias is related to accuracy in that the term represents a lack of agreement between an actual value and the value obtained in an analytical or testing procedure. Bias is a constant, determinate error present in the system, rather than an indeterminate or random error. Biases are caused by method inconsistencies or deficiencies in the measurement system.

Assumptions

The use of control charts, as well as other statistical techniques, makes several assumptions regarding the process or procedure to which it is applied. The first assumption is that there will be variation in the data. No process or procedure is so perfect or so unaffected by its environment and other variable factors that it will always give exactly the same assay value or product. Where such situations seem to exist, either the device used to measure the process is not sensitive enough or the person making the measurements is not performing properly (Figure 4.1).

The sources of variation that will be present in an analytical procedure will include: 1) differences among analysts, 2) differences among instruments, 3) differences among reagents and related supplies, 4) differences of 1, 2, and 3 over a period of time, and 5) differences of 1, 2, and 3 with each other and with time.

For testing procedures, the sources of variation will be the same, except for 3.

Since variation in results must be accepted as a way of life, its assessment and control become highly important. A stable "system of chance causes" is inherent in the nature of any process or procedure, and this "system of chance causes" will produce a pattern of variation.

When this pattern of variation is stable, the process or procedure is said to be "in statistical control" or just "in control." Any variation outside this pattern will have an assignable cause, which can be determined and corrected.

The presence of a "stable system of chance causes" allows one to apply the laws of probability to the analysis of the data. Statistical analysis techniques assume the randomness of the data (our second assumption) and are valid only in such a system.

The third assumption is that sample averages will be normally distributed, as seen in the last chapter.

The control chart technique provides a means of separating a stable pattern of variation from assignable cause variations. The control chart provides a graphical representation of the process or method test data so that the variability of all results is compared with the average or expected variability with small, arbitrarily defined samples or subgroups of the data. The control chart then compares "within groups" variability to "between group" variability. The technique is, in effect, a graphical "analysis of variance."

The data is plotted on the vertical scale in units of the test or analytical results, and on the horizontal scale in units of time or sequence of results. The average value or mean and limits on the degree of dispersion or spread of the results are calculated and entered on the chart. The result is the control chart depicted in Figure 4.2.

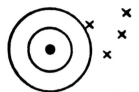

IMPRECISE AND INACCURATE

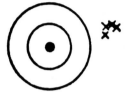

PRECISE BUT INACCURATE

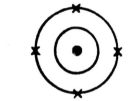

ACCURATE BUT IMPRECISE

PRECISE AND ACCURATE

FIGURE 4.1.

Determining appropriate control limits is a complex subject. Control limits may be based on the capability of the process itself and calculated from the observations made. It is common practice to set the control limits at $\pm 3\sigma$ above and below the mean. Since the distribution of averages (means) tends to be normal in form, the probability of getting values outside the control limits can be readily calculated. Non-normal distributions will differ in percentage of values beyond the $+\pm 3$ σ

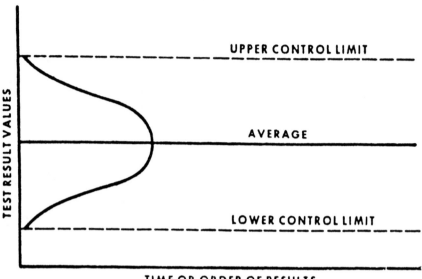

FIGURE 4.2.   Control Chart.

limits. In laboratory work, it is often desirable to establish upper and lower *warning* limits, as well as control limits. Warning limits are commonly placed at $\pm 2\sigma$ above and below the mean.

The control chart is, then, a graphical presentation of the observed results plotted in relation to calculated limits, generally in the order of occurrence or sequentially in time. If the process is "in control," the forthcoming results should fall within the calculated control limits to the degree expected. There also should not be any occurrence of trends, cycles, runs above or below the mean, or runs up or down.

When these conditions are met, the analyst or tester has a degree of assurance that no drastic change has taken place in his procedure. That is, no assignable cause has disturbed the state of "statistical control."

If the tester were to use the mean of several values as a single control chart point (an X chart) the chances of picking up small changes in the process average are increased. The protection against not detecting small changes in the process average increases as the sample size grows larger.

The choice of the sample or subgroup size is, therefore, very important. Shewhart suggested four as the ideal subgroup size, but sampling limitations often force the laboratory worker to settle for samples of two. In manufacturing statistical process control, five is often used as the sample size, since samples are usually readily available and the sample mean is easy to calculate by multiplying the sample sum by .20. Shewhart control charts ($\overline{X}$) were usually set up with a sample size of n = 4. The use of frequent sampling will also detect changes more quickly with time. The ultimate in control would be to use large samples taken frequently. However, an economic decision must be made, weighing the value of closer control versus the increased cost of attaining that degree of control.

CONTROL CHART PROCEDURES

The following is a step-by-step explanation of how to set up, initiate, and maintain an $\overline{X}$-R control chart, on which both the sample means and the ranges are plotted. The details of how to make the calculations and plot the data will be explained in a later paragraph.

There are a number of decisions to be made and actions to be taken in setting up a control chart for a given procedure. These steps are outlined below.

Decisions preliminary to setting up the control chart:

1. Decide upon the objective of the control chart—what is being controlled.
2. Choose the variable to be measured. There are, of course, a myriad of variables developed during the testing and analytical activities of a laboratory. The problem is to select candidates for control charting that will yield the most benefit for the effort spent on setting up, maintaining, and interpreting the chart. It is obvious that it is impractical to control every method used in the laboratory, so laboratory policy should be to use charting for only the most troublesome spots in laboratory operations.

   Typical variables that might be the subject of control charting are:
   a. Rockwell "C" hardness, in lots of a steel part after heat treat,
   b. Moisture content of molding sand in a brass foundry;
   c. Percent manganese in samples of steel, from successive heats;
   d. Tensile strength of SAE 1010 carbon steel, from successive heats;
   e. Gas flow resistance for filter media;
   f. Lubricating oil fill volumes at a bottling operation; and
   g. Carbon monoxide monitoring results from the same location on successive days.
3. Decide upon the basis for subgrouping. (A word of explanation as to the kind of decision to be made is in order here. The basic function of the control chart is to compare "within group" variability with "between group" variability. For a single analyst or tester running a procedure, the "within group" data may represent one day's output, and the "between group" data may represent the between days or day-to-day variability. In the case where several different operators, instruments, or laboratories are involved, selecting the subgroup unit is critical. Assignable causes of variation should show up as "between

group" and not as "within group" variability. Thus, if the workers are thought to be the sources of assignable causes of variation, their results should not be lumped together in a "within group" subgrouping.)
4. Decide on the size of the subgroup based on data available and economic considerations.
5. Choose the frequency of subgroup sampling.
6. Set up the $\overline{X}$-R worksheet forms.
7. Choose the method of measurement. (Note: This last step may often be dictated by the method and may not be a decision at all.)

Initiating the Control Chart

1. Make the measurements.
2. Record the measurements and all relevant data on the work sheet.
3. Calculate the average, $\overline{X}$, for each subgroup.
4. Calculate the range, R, for each subgroup
5. Plot the data on the $\overline{X}$ chart.
6. Plot the data of the R chart.

Determine Trial Control Limits

1. Decide on the number of subgroups needed, in order to calculate control limits.
2. Calculate average range, $\overline{R}$.
3. Calculate the upper and lower control limits for R.
4. Calculate the average of $\overline{X}$ values, to arrive at the grand mean, $\overline{\overline{X}}$.
5. Calculate the upper and lower control limits for $\overline{X}$.
6. Plot the $\overline{X}$ and R lines and their corresponding control limits on the appropriate control charts. (Note: Conventional usage is to use solid lines to plot the $\overline{X}$ and $\overline{R}$ lines and dashed lines to indicate control limits.)

The number of subgroups necessary to calculate control limits for subgroups of 4 is 25. However, in testing and analytical work, it is common, because of the scarcity of data, to use preliminary control limits that are based on fewer data points. Of course, any action prompted by such a control chart would have to be weighed against the quantity of available data.

Draw Preliminary Conclusions

1. Does the chart indicate an "in control" state?
2. Is the method performing to expectations?
3. What actions, if any, are prompted by the control chart?

If the control charts indicate that the procedure is in a state of control and conforms to expectations, the analyst or tester can use the control chart to monitor and maintain the desired quality level of the procedure. If the procedure is "out of control" or does not meet expectations, corrective action is indicated.

Routine Operation

1. Post new data to the charts in a timely manner, so that they reflect the nature of the operation on a real-time basis.
2. Use the control chart to maintain control.
3. Reevaluate periodically.
4. Recalculate and replot control limits, as necessary.

CALCULATIONS

The $\overline{X}$ and R charts are, perhaps, the most common form of control chart used in laboratories. The use of $\overline{X}$, rather than individual X's (data points based on single observations), offers two advantages. As already indicated, a larger subgroup or sample size, n = 4 or larger, will detect smaller changes in the process average than the use of smaller subgroups. Using $\overline{X}$ also offers protection against non-normal distributions of the basic data. Shewhart proved that in sampling from either rectangular or right triangular distributions—extremely non-normal distributions—a sample size of n = 4 gave close to a normal distribution for the observed $\overline{X}$'s. Since the distribution of data conforming to the normal curve can be completely specified using $\overline{X}$ and σ, the use of $\overline{X}$ charts simplifies and makes more exact the calculation of control limits.

Using the range R as an estimate of the standard deviation, rather than the standard deviation, σ, is justified for small sets of data (n $\leq$ 10), since it is almost as efficient statistically and provides an easier method of calculation.

CALCULATION OF CONTROL LIMITS

We will skip over the computations for $\overline{X}$ and R, since they were covered in Chapter 2, and go directly to the calculations for control limits, using the range data and factors that have been developed for this purpose, reducing the task to a simple multiplication process.

The equations needed to acquire the data needed for constructing the $\overline{X}$-R chart appear in Table 4.1, where:

1. k = number of subgroups.
2. Factors A_2, D_4, and D_3 are obtained from Table 4.2.

TABLE 4.1.   Equations for Control Chart Computations.

Grand Average	$\overline{\overline{X}} = \dfrac{\Sigma \overline{X}}{k}$	(1)
Control Limits on Average	$CL = \overline{\overline{X}} \pm A_2 \overline{R}$	(2)
Range	$\overline{R} = \dfrac{\Sigma R}{k}$	(3)
Upper Control Limit on Range	$UCL_R = D_4\overline{R}$	(4)
Lower Control Limit on Range	$LCL_R = D_3\overline{R}$	(5)

The values given in Table 4.2, Factors for Computing Control Chart Limits, enable the tester to calculate the control limits for the procedure. Factor D_2, when multiplied by the average range, $\overline{R}$, provides an easily computed estimate of the sample standard deviation, s. Using factor D_2 will be discussed in detail in Chapter 6. Note that the lower control limit on $\overline{R}$ is zero when $n \leq 6$.

CALCULATION OF WARNING LIMITS

The control limits plotted include approximately the entire set of data, when under "in control" conditions, as the limits correspond to ± 3 sigma limits, which, as seen earlier, encompass 99.7% of all the data. It is sometimes desirable, however, in laboratory work, to take milder action when results exceed the ± 2 sigma limits, where about 95% of the results are expected to fall. Such limits are called "warning limits." The "Upper Warning Limit (UWL)", for ranges, is related to the mean range ($\overline{R}$). Thus, the UWL can be computed by using the following equations:

$$UCL = D_4\overline{R} \qquad (4\text{-}1)$$

TABLE 4.2.   Factors for Computing Chart Lines.

Observations in Subgroup (n)	Factor A_2	Factor d_2	Factor D_4	Factor D_3
2	1.88	1.13	3.27	0
3	1.02	1.69	2.58	0
4	0.73	2.06	2.28	0
5	0.58	2.33	2.12	0
6	0.48	2.53	2.00	0
7	0.42	2.70	1.92	0.08
8	0.37	2.85	1.86	0.14

$$UWL = 2/3 \ (D_4\overline{R} - \overline{R}) + \overline{R} \qquad (4\text{-}2)$$

Appropriate warning limits are constructed for the $\overline{X}$ chart by using two-thirds of the values found for the upper and lower control limits on $\overline{X}$.

EXAMPLE—CONTROL CHART
FOR REAGENT BLANK DETERMINATIONS

The data shown in Table 4.3 gives the results from a laboratory running blank determinations, as a normal practice at the beginning and end of each day's run. The data is taken from the results posted in an actual laboratory notebook. The readings represent the observed percent transmission with a 1.0 μg Hg reference standard sample being used. The records show that these tests were run over a period of time that ran from August until the end of December, in one year. Such slow accumulation of data is often the case in laboratory practice. Since there are only 8 sets of data available, any control limits calculated would be very preliminary in nature, since normally, 20 or more sets of data (subgroups) are recommended for setting up control limits. The calculations involved in setting up the control chart for this

TABLE 4.3.   Mercury Quality Control Data

Date	No.	Blank Determinations (%)	
9 Aug	10	AM	97.0
9 Aug	10	PM	97.1
23 Aug	11	AM	97.1
23 Aug	11	PM	97.5
4 Sep	12	AM	—
4 Sep	12	PM	98.2
26 Sep	13	AM	98.0
26 Sep	13	PM	98.9
12 Oct	14	AM	99.0
12 Oct	14	PM	99.5
31 Oct	15	AM	99.0
31 Oct	15	PM	98.9
6 Nov	16	AM	99.0
6 Nov	16	PM	98.5
26 Nov	17	AM	99.0
26 Nov	17	PM	97.8
3 Dec	18	AM	99.0
3 Dec	18	PM	97.8
7 Dec	19	AM	99.5
7 Dec	19	PM	—
21 Dec	20	AM	100.0
21 Dec	20	PM	100.0

Laboratory _XYZ Laboratory Inc_   **Date** _Aug - Dec_

Method of Test or Operation _Hg / Blank Determination_

Reference Value _Blank (0 Added)_   **Increment of Measurement** _% T_

Data

Date	No.	X_1	X_2	X_3	$\bar{X}$	R
	1	97.0	97.1		97.05	0.1
	2	97.5	-			
	3	98.2	98.0		98.1	0.2
	4	98.9	99.0		98.95	0.1
	5	99.5	99.0		99.25	0.5
	6	98.9	99.0		98.95	0.1
	7	98.5	99.0		98.75	0.5
	8	99.0	97.8		98.4	1.2
√	9	99.5	-			
	10	100.0	100.0		100.0	0
	11					
	12					
	13					
	14					
	15					
	16					
	17					
	18					
	19					
	20					

Totals $\Sigma \bar{X}$ _789.45_ ΣR _2.7_

X_1 = observed value   R = largest − smallest
k = sets of values   CL = control limit
Σ = summation   WL = warning limit
U = upper   L = lower
D_4 = 3.267 for n'= 2; 2.575 for n'= 3
A_2 = 1.880 for n'= 2; 1.023 for n'= 3
n' = number of values in the set

Calculations

1. $\bar{R}$ = $\Sigma R \div k$
   0.34 = 2.7 ÷ 8

2. UCL_R = $D_4 \times \bar{R}$
   1.11 = 3.268 × 0.34

3. UWL_R = $2/3(D_4\bar{R} - \bar{R}) + \bar{R}$
   0.86 = 2/3(1.11 - .34) + .34

4. $\bar{\bar{X}}$ = $\Sigma \bar{X} \div k$
   93.68 = 789.45 ÷ 8

5. $CL_{\bar{X}}$ = $A_2 \times \bar{R}$
   0.64 = 1.88 × 0.34

6. $WL_{\bar{X}}$ = $2/3 \times CL_{\bar{X}}$
   0.43 = 2/3 × 0.64

7. $UCL_{\bar{X}}$ = $\bar{\bar{X}} + CL_{\bar{X}}$
   99.32 = 98.68 + 0.64

8. $UWL_{\bar{X}}$ = $\bar{\bar{X}} + WL_{\bar{X}}$
   99.11 = 98.68 + 0.43

9. $LWL_{\bar{X}}$ = $\bar{\bar{X}} - WL_{\bar{X}}$
   98.25 = 98.68 - 0.43

10. $LCL_{\bar{X}}$ = $\bar{\bar{X}} - CL_{\bar{X}}$
    98.04 = 98.68 - 0.64

FIGURE 4.3. $\bar{X}$-R Chart Calculation Worksheet.
Laboratory Quality Control.

analytical situation are listed along the right-hand side of the work sheet shown in Figure 4.3. When completed, the results of the calculations are transferred to the bottom of the X̄-R Chart worksheet, Figure 4.4, and then are plotted on the charts, in accordance with the directions appearing at the bottom of the worksheet.

The control chart is then examined, to note any out-of-control points or any runs up or down or above or below the grand mean line. In this case, since the

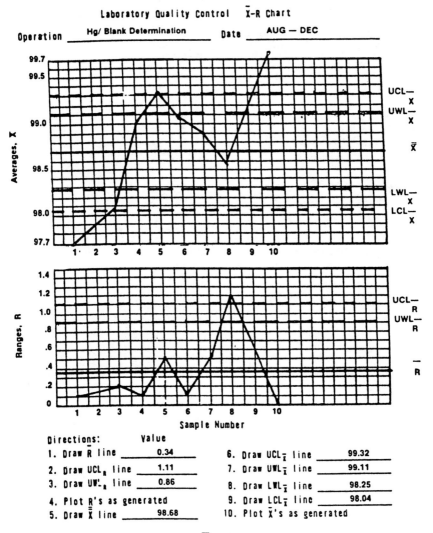

Directions:

	Value			Value
1. Draw R̄ line	0.34	6. Draw UCL$_{\bar{x}}$ line	99.32
2. Draw UCL$_R$ line	1.11	7. Draw UWL$_{\bar{x}}$ line	99.11
3. Draw UWL$_R$ line	0.86	8. Draw LWL$_{\bar{x}}$ line	98.25
4. Plot R's as generated		9. Draw LCL$_{\bar{x}}$ line	98.04
5. Draw X̿ line	98.68	10. Plot X̄'s as generated	

FIGURE 4.4. Laboratory Quality Control X̄-R Chart.

out-of-control points have occurred in the past, no immediate action to determine their cause may be appropriate. Any future out-of-control sets of data should be studied, to determine if there is an assignable cause. Once determined, corrective action should be taken, to remove any assignable cause.

All new data sets should be recorded on the worksheets as they are generated. This way the control chart represents the real-time situation. This allows for ongoing analysis and feedback for prompt corrective action, if necessary. The appropriate actions in response to out-of-control points are analytical or technical actions and are not statistical in nature. The skills of the analyst or technologist will enable him to determine and resolve the technical problem that caused the statistical anomaly to occur.

References

Shewhart, W. A. 1931. *Economic Control of Manufactured Product.* Princeton, NJ: Van Nostrand Reinhold.

1983. *Industrial Hygiene Laboratory Quality Control.* Cincinnati, OH: National Institute for Occupational Safety and Health.

5

Special Situation Control Charts

It sometimes happens that the nature of the available data is such that $\overline{X}$-R charts cannot be used with that data. In these cases, other types of control charts may be used or other techniques employed, to plot the data in a manner similar to that appearing in $\overline{X}$-R chart plots.

CONTROL CHARTS FOR INDIVIDUAL VALUES

In many instances, a rational basis for subgrouping may not be available, or the test or analysis may be so infrequent that it requires action on the basis of individual results. In such cases, X charts for individual values are used. However, the control limits must come from some source, in order to obtain a measure of "within group" variability. Using specification or tolerance limits is appropriate in these cases (Figure 5.1).

There are certain disadvantages associated with using control charts for individuals, which must be recognized:

1. The chart is as responsive as changes in the average.
2. Changes in dispersion are not detected unless an R chart, based on some subgrouping, is used.
3. The distribution of the raw data must be approximately normal, for the control limits to be valid.

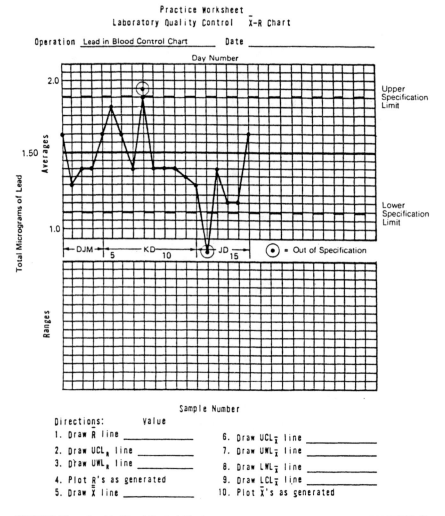

FIGURE 5.1.   Lead in Blood Control Chart.

MOVING AVERAGES AND RANGES

The $\overline{X}$ control chart is more efficient than an X chart for disclosing changes in the average as the subgroup size increases. A logical compromise between the $\overline{X}$ and the X approach would be applying the moving average, which is plotted for a given set of data. To illustrate, the moving average is computed as follows for the data given in Table 5.1.

TABLE 5.1.   Moving Average and Range Table (n-2).

Sample No.	Assay Value	Sample Nos. Included	Moving Average	Moving Range
1	17.09			
2	17.35	1-2	17.22	0.26
3	17.40	2-3	17.38	0.05
4	17.23	3-4	17.32	0.17
5	17.09	4-5	17.16	.14
6	16.94	5-6	17.02	0.15
7	16.68	6-7	16.81	0.26
8	17.11	7-8	16.90	0.43
9	18.47	8-9	17.79	1.36
10	17.08	10-11	17.08	1.39
11	17.08	10-11	17.08	0.00
12	16.92	11-12	17.00	0.16
13	18.03	12-13	17.48	1.11
14	16.81	13-14	17.42	1.22
15	17.18	14-15	16.98	0.37
16	17.34	15-16	17.25	0.16
17	16.71	16-17	17.03	0.63
18	17.28	17-18	17.00	0.57
19	16.54	18-19	16.91	0.74
20	17.30	19-20	16.92	0.76

$$\overline{X} = 17.18 \qquad \overline{R} = .52$$

First the average for samples 1 and 2 is calculated. Then sample 1 is dropped and sample 3 is picked up and that average is computed. This process of dropping the trailing data point and picking up the next one is continued for the whole data set, thus yielding a series of averages, rather than just one X line.

The moving range is computed by subtracting the smaller from the larger of succeeding samples. Thus, the moving range for sample 2 is 17.35 - 17.09 = 0.26, and so forth. The moving range is a good measure of acceptable variation when no rational basis for subgrouping is available or when results are infrequent or expensive to gather. The $\overline{X}$ and R can be used to compute control limits in a manner similar to those of the $\overline{X}$-R charts, but their interpretation and use is not the same. Therefore, care should be used when determining corrective action based on the occurrence of runs. For a discussion on how to interpret runs, see Chapter 12.

OTHER CONTROL CHARTS FOR VARIABLES

Although the standard $\overline{X}$ and R chart for variables is the most commonly used, it does not always do the best job. Several examples follow, showing where other charts are more applicable.

Variable Subgroup Sizes

As seen in the previous chapter, the standard $\overline{X}$ and R chart is applicable for a constant size subgroup of n = 2, 3, 4, or 5. In some cases, however, consistently equal subgroup sizes are not available, due to the nature of the sampling scheme or sample availability. When this happens, control limit values must be calculated for each subgroup separately. The results are plotted in the usual manner, with the control limits ruled in for each subgroup individually. One way to avoid burdensome calculations is to obtain a weighted mean and standard deviation for the combined groups; another is to use an average subgroup size.

R or Sigma (σ) Charts

In some situations, such as when using the results of chemical analyses, the dispersion or variability of the data is equal over a range of assay values. In such cases, a control chart for either the range or the standard deviation is appropriate, but a mean chart is not. When the dispersion is a function of the concentration, control limits can be expressed in terms of a percentage of the mean. In practice, such control limits would be given as shown below:

± 5 units/liter for 0-100 units/liter concentration
± 5% for 100 units/liter concentration

The sigma (σ) chart is also appropriate when the subgroup is large enough to estimate the standard deviation from its range, or when sigma can be calculated from the raw data.

Cumulative Sum Control Charts

Cumulative Sum (CuSum) control charts provide a running, visual summation of deviation, from some preselected reference point. While Shewart control charts are useful in controlling single out-of-control points, CumSum charts are more effective in identifying gradually encroaching out-of-control conditions, such as those caused by the gradual degradation of stock standard solutions used for calibration, or by the slow degradation of instrument performance.

We will discuss, as an example, using a CuSum chart for duplicate samples. First, consideration must be given to the number of duplicate analyses to be conducted in a series of samples. Likewise, the same decision must be made on spiked or standard samples.

In considering the number of duplicate sample analyses to be conducted in a series of samples, it is necessary to weigh the consequences when the data go out of control. When this happens, the alternatives are to reanalyze a series of samples or to discard the questionable data obtained. The samples to be reanalyzed are those lying between the last in-control data point and the out-of-control point presented. A realistic frequency for running duplicate samples would be every fifth sample; however, economic considerations and experience may require more or less frequent duplicate sample analyses. Once the frequency of duplicate samples has been determined, it is then necessary to prepare standard samples in concentrations relative to the concentration of the control charts. These samples should be similar to those of the analytical samples. The standard samples must be intermittently dispersed among the series of samples to be analyzed, without the analyst's knowledge of concentration. Similarly, duplicate samples must also be scattered throughout the series of samples, again, ideally, without the analyst's knowledge; however, this is sometimes very difficult to accomplish.

The results of the duplicate and spiked sample analyses should be calculated as soon as the analysis of the samples is completed, to allow for early detection of problems that may exist in the laboratory. An example of these calculations follows:

Duplicate Sample No. M	Results No. 1	No. 2	Difference (di)	di^2	$\Sigma(di^2)$
1	5.4	5.2	.2	.04	.04
2	4.8	4.7	.1	.01	.05
3	6.1	5.8	.3	.09	.14

Where di = the difference between the i^h set of duplicates or spiked samples. Upon plotting the summation or $\Sigma(di^2)$, three possibilities can occur (Figure 5.2). For detailed instructions on how to plot a CuSum chart, see pp. 24-26 to 24-29 in Juran's *Quality Control Handbook*, 4th Ed.

1. Out of control on the upper limit. When data goes out of control on the upper limit, the following steps should be taken:
   a. Stop work immediately.
   b. Determine problems:
      (1) Precision control chart
         (b) The analyst
         (b) Nature of the sample
         (c) Glassware contamination
      (2) Accuracy control chart
         (a) The analyst
         (b) Glassware contamination

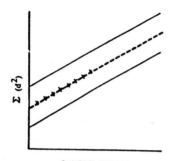

SAMPLE SET NO.

ANALYSIS IN CONTROL

NO PROBLEMS:
   CONTINUE ANALYSIS

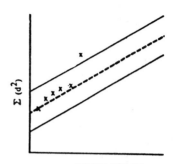

SAMPLE SET NO.

ANALYSIS OUT OF CONTROL
   UPPER LIMIT

PROCEDURES:
   1. STOP ANALYSIS
   2. LOCATE PROBLEM
   3. CORRECT PROBLEM
   4. RERUN SAMPLES
   5. START CHART AT SAMPLE
      SET NO. 1.

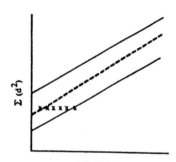

SAMPLE SET NO.

ANALYSIS OUT OF CONTROL
   LOWER LIMIT

INCREASED EFFICIENCY OR
FALSE REPORTING
PROCEDURES:
   1. CONTINUE ANALYSIS
   2. CONSTRUCT NEW CHART
      WITH RECENT DATA
   3. OBSERVE ANALYST

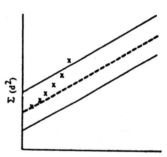

SAMPLE SET NO.

ANALYSIS OUT OF CONTROL
   UPPER LIMIT

CONTINUOUS ERROR TREND
PROCEDURES:
   SAME AS ABOVE BUT STOP
   ANALYSIS WHEN TREND IS
   DETECTED.

FIGURE 5.2.   Laboratory Quality Control Charts.

      (c) Contaminated reagents

      (d) Instrument problems

      (e) Sample interference with spiked material

  c. Rerun samples represented by that sample set number, including additional duplicate and spiked samples.

  d. Begin plotting at sample No. 1 on chart.

2. In control within the upper and lower limit lines—When data continuously fall in between the upper and lower control limits, the analyses should be continued until an out-of-control trend is detected.

3. Out of control on the lower limit—When data fall out of control on the lower limit, the following steps should be taken:

  a. Continue analyses, unless trend changes.

  b. Construct new control charts, based on most recent data.

  c. Check analyst's reporting of data.

Fraction Defective Charts (p Charts)

When testing equipment or when accessories are purchased in large numbers, as may be the case with personal dust sample filters, it may be desirable to record the percentages of defectives found in weighed samples, drawn from each lot. Defectives are considered to be those filters not meeting the manufacturer's indicated weights. In this case, the percentages of defectives, p, are plotted. If the sample sizes are constant, np may be plotted. p is obtained by dividing the number of defectives found by the sample size n. The actual number of defectives is then represented by np, the quantity that is divided by n, yields p. Thus:

$$p = np/n \qquad (5\text{-}1)$$

The p's, or percent defective data points, are plotted in the same manner as $\overline{X}$'s are plotted on $\overline{X}$-R control charts. The control limits are plotted at ± three standard deviations above and below the mean, using the equation:

$$\sigma_{p'} = \sqrt{\frac{p'(1 - p')}{N}} \qquad (5\text{-}2)$$

to calculate the standard deviation, where p' represents the fraction defective. The chart is interpreted in the same way as $\overline{X}$-R charts.

Although this form of chart is another variation of the Shewhart control chart, the p chart is not a substitute for Shewhart variables control charts, since it differs in two important aspects. First, the p chart requires, in general, larger sample sizes than the X-bar, R chart, in order to establish percentages. Secondly, the variables chart deals with measurable observations that can be recorded in terms of numerical quantities, while the percentages are based on go-no-go attributes. Examples are:

The filter meets the specified weight, or it does not.
The steel has a hardness of Rockwell C, or it does not.
The detector tubes are .275″ diameter maximum, or they are not.

In many cases, attributes are not easily quantifiable, as in the case of discolored bulk materials or other cosmetic characteristics.

6

Statistical Tools
for the Laboratory

INTRODUCTION

When handling the large amounts of data generated in the laboratory, questions often arise that require answers not afforded by control charts. This chapter explains how to use some of these techniques and how to employ the statistical tables associated with them. Examples are provided in "cookbook" fashion, to provide one ready-reference source for this useful information.

TESTS FOR SIGNIFICANCE
OF DIFFERENCES

There is often the need to determine whether or not differences in results between different sets of data, although appearing to agree, are really statistically different. The solution to this problem depends upon the kind and amount of information that is available in each circumstance. The following table shows the kind of information needed to employ the test selected, whether it be the t, Chi-square ($_\kappa 2$), or F test. (Table 6.1.)

Note that the power of these tests is dependent on sample sizes. When samples are too small to afford a powerful test, or there are large variabilities, the result may be that an important difference does exist that is, however, not statistically significant. Conversely, there may be small variability and large sample sizes that afford a very powerful test—so powerful that unimportant differences are found to be significant.

Table 6.1   Tests for Significance of Difference

Test No.	Type	Is there a Significant Difference Between	When
1.	t	The sample mean $\overline{X}$, and the population mean $\overline{X}'^{,*}$	The population standard deviation is known. (σ)
2.	t	The sample mean $\overline{X}$, mean and the population mean $\overline{X}'$	The population standard deviation is unknown and estimated from the sample. (s)
3.	t	Two sample means: $\overline{X}_1$ and $\overline{X}_2$	The population standard deviation (σ) is known and is the same for both samples.
4.	t	Two sample means: $\overline{X}_1$ and $\overline{X}_2$	The population standard deviation (σ) is not known but is believed to be the same for both.
5.	χ^2	The sample variability and the population variability	The population standard deviation (σ) is known.
6.	F	Two sample variabilities σ_a and σ_b	The population standard deviation is unknown.

NB:* In this chapter, we will follow the notation of Juran (1962) and use $\overline{X}'$ to represent the population mean, since these examples were adapted from that work.

Adapted by permission from Quality Control Handbook, edited by J. M. Juran, 2nd Ed. Copyright 1962 by McGraw-Hill, Inc. Used with permission of McGraw-Hill Book Co.

1.   The "t" Test for Difference Between a Sample Mean ($\overline{X}$) and a Population Mean ($\overline{X}'$)[1]

This form of the "t" test is applied when the population standard deviation (σ) is known. The following example deals with the case of a large number of urine samples checked for mercury content. Past tests have indicated an average concen-

Specimen No.	μg Hg/Liter
1	5.02
2	4.87
3	4.95
4	4.88
5	5.10
6	4.93
7	4.91
8	5.09
10	4.89
11	5.06
12	4.85
Total	59.42
Average $\overline{X}_1$ =	4.95

---

[1]Adapted by permission from Quality Control Handbook, edited by J. M. Juran, 2nd Ed. Copyright 1962 by McGraw-Hill, Inc. Used with permission of McGraw-Hill Book Co.

tration of 5.15 µg mercury/liter of urine and a standard deviation 0.25 µg/l. Later, during the most recent series of tests, a set of samples from the same individuals was checked, with the following results:

The question posed, then, is: Is the last sample average, 4.95 µg Hg/liter, significantly different from that of the previous samples. 5.15 = µg Hg/liter?

Solution: The difference between past performance and the average of the sample from the new round of tests is:

$$5.15 - 4.95 = 0.20$$

To interpret the significance of this difference, use is made of the fact that the natural pattern of the averages coming from a controlled environment is well known and can be predicted from past variability (σ). The observed difference is therefore compared with the expected pattern as a means of determining how unusual the observed difference is by finding the value of the ratio:

$$t = \frac{\overline{X} - \overline{X}'}{\dfrac{\sigma}{\sqrt{n}}}$$

(6-1)

Substituting the proper values:

$$t = \frac{5.15 - 4.95}{0.25/\sqrt{12}}$$

$$t = 2.77$$

To determine how significant this value is, it is necessary to consult a table of "t" values (Table 6.2 "Distribution of t"). When the value of σ is well established, as in this case, the table of "t" values is entered on the lowest line of the table, DF = ∞. Find, among the values of "t" given on the bottom line, the one closest to 2.77. We find that 2.77 falls between 2.576 and 3.291, which lie within the columns that indicate the probabilities of 0.01 and 0.001, respectively. Interpolation is not necessary in this case, because it is obvious that the probability of having so great a difference between the two averages, by chance alone, is considerably less than 0.01.

Conclusion:

The conclusion in answer to the question posed is that the difference in the past averages and the average results of the newest samples is significant, because the odds of 100 to 1 of this occurring are too great to be attributed to sampling variation alone.

Table 6.2   * Distribution of *t*

Value of *t* corressponding t certain selected probabilities (i. e., tail areas under the curve). To illustrate: the probability is 0.975 that a sample with 20 degrees of freedom would have *t.* = + 2.086 or smaller.

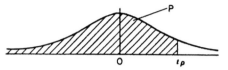

DF	$t_{.60}$	$t_{.70}$	$t_{.80}$	$t_{.90}$	$t_{.95}$	$t_{.975}$	$t_{.99}$	$t_{.995}$
1	0.325	0.727	1.376	3.078	6.314	12.706	31.821	63.657
2	0.289	0.617	1.061	1.886	2.290	4.303	6.965	9.925
3	0.277	0.584	0.978	1.638	2.353	3.182	4.541	5.841
4	0.271	o.569	0.941	1.476	2.015	2.571	3.365	4.032
5	0.267	0.559	0.920	1.476	2.015	2.571	3.365	4.032
6	0.265	0.553	0.906	1.440	1.943	2.447	3.143	3.707
7	0.263	0.549	0.896	1.415	1.895	2.365	2.998	3.499
8	0.262	0.546	0.889	1.397	1.860	2.306	2.896	3.355
9	0.261	0.543	0.883	1.383	1.833	2.2262	2.821	3.250
10	0.260	0.542	0.879	1.372	1.812	2.228	2.764	3.169
11	0.260	0.540	0.97	1.363	1.796	2.2201	2.718	3.106
12	0.259	0.539	0.873	1.356	1.782	2.179	2.681	3.055
13	0.259	0.538	0.870	1.350	1.771	2.160	2.650	3.012
14	0.258	0.537	0.868	1.345	1.761	2.145	2.624	2.977
15	0.258	0.536	0.866	1.341	1.753	2.131	2.602	2.947
16	0.258	0.535	0.865	1.337	1.746	2.120	2.583	2.921
17	0.257	0.534	0.863	1.333	1.740	2.110	2.567	2.898
18	0.257	0.534	0.862	1.330	1.734	2.101	2.552	2.878
19	0.257	0.533	0.861	1.328	1.729	2.093	2.539	2.861
20	0.257	0.533	0.860	1.325	1.725	2.086	2.528	2.845
21	0.257	0.532	0.859	1.323	1.721	2.080	2.518	2.831
22	0.256	0.532	0.858	1.321	1.717	2.074	2.508	2.819
23	0.256	0.532	0.858	1.319	1.714	2.069	2.500	2.807
24	0.256	0.531	0.857	1.318	1.711	2.064	2.492	2.797
25	0.256	0.531	0.860	1.325	1.725	2.086	2.528	2.845
26	0.256	0.531	0.856	1.315	1.706	2.056	2.479	2.779
27	0.256	0.531	0.855	1.314	1.703	2.052	2.473	2.771
28	0.256	0.530	0.855	1.313	1.701	2.048	2.467	2.763
29	0.256	0.530	0.854	1.311	1.699	2.045	2.462	2.756
30	0.256	0.530	0.854	1.310	1.697	2.042	2.457	2.750
40	0.255	0.529	0.851	1.303	1.684	2.021	2.423	2.704
60	0.254	0.527	0.848	1.296	1.671	2.000	2.390	2.660
120	0.254	0.526	0.845	1.289	1.658	1.980	2.358	2.617
∞	0.253	0.524	0.842	1.282	1.645	1.960	2.326	2.576

*Adapted by permission from W.J. Dixon and F. J. Massey, Jr., "Introduction to Statistical Analysis," 3d cd., McGraw-Hill Book Company, New York, copyright © 1969. Entries originally from Table III of R. A. Fisher and F. Yates, "Statistical Tables," Oliver & Boyd, Ltd., London.

2. The "t" Test For the Difference Between the Sample Mean (X), and the Population Mean (X')[2]

This form of the "t" test is applied when the population standard deviation (σ') is unknown and must be estimated from the sample.

The following example deals with records of tests for carbon monoxide that indicate that the average amount found in a certain plant location, over the period of a year, was 49.95 pp.—No record of individual measurements was kept. After a change was made in plant operations, 61 new samples were taken, and an analysis showed an average of 54.62 ppm of CO and a standard deviation of 5.34.

Procedure	Example

1. Estimate the standard deviation from the sample:

$$\sigma' = \sqrt{\dfrac{\Sigma(X - \overline{X})^2}{n - 1}}$$

1. $\sigma' = 5.34$

(6-2)

2. Compute the ratio:

2. $t = \dfrac{54.62 - 49.95}{5.34/\sqrt{61}}$

$$t = \dfrac{\overline{X} - \overline{X}'}{\sigma'/\sqrt{n}}$$

$t = 6.78$ (6-3)

3. Enter the table of t values (Table 6-2) at DF = 60 (n-1) and the selected confidence.

3. 6.78 from step 2 is larger than any value given for DF = 60.

4. If the computed ratio for t (6.78) is larger han t (table value from step 3), decide that the average of the new samples is different from that of the standard; otherwise, there is no reason to believe that they differ.

4. Conclude that the last samples taken are significantly higher in carbon monoxide, since a change difference in average percentage 54.62 - 49.95 = 4.67 is so unlikely.

The question to be answered is: Is the average of carbon monoxide present significantly different, under the new operating conditions,than it was before?

---

[2] Adapted by permission from *Quality Control Handbook*, edited by J. M. Juran, 2nd. Ed. Copyright 1962 by McGraw-Hill, Inc. Used with permission of McGraw-Hill Book Co.

3. The "t" Test For the Difference Between Two Sample Means ($\overline{X}_1$ and $\overline{X}_2$), Based on Independent Random samples.[3]

This form of the "t" test applies when the σ' is known and is the same for both sets of samples.[4]

Here we have the case of a a manufacturer of automobile batteries, who has two plants. Eight blood samples are drawn from employees in similar production areas in each plant, with the results shown below. The population standard deviation has been estimated, based on long experience, to be 0.22 σg Pb/100g blood. Is there a significant difference in the exposure to lead, between the two plants?

Plant A σgPb/100g Blood   Plant B σgPb/100g Blood

Plant A	Plant B
12.53	12.53
12.37	13.22
12.48	13.01
12.77	12.97
12.52	12.96
12.81	13.03
12.76	12.82
12.52	13.43

Totals   100.76   103.97
$\overline{X}_1 = 12.60$ $\overline{X}_2 = 13.00$

Procedure	Example

1. There is no prior knowledge of the population value corresponding to the measure being tested. The difference under study is between two quantities, each of which has some sampling error. The error contributed by each of the samples is inversely proportional to the sample size.

1. $t = \dfrac{x_1 - x_2}{\sigma\sqrt{(n_1 + n_2)/n_1 n_2}}$ (6-4)

2. Compute:

$$t = \frac{X_1 - X_2}{\sigma\sqrt{\dfrac{(n_1 + n_2)}{n_1 n_2}}}$$

2. $t = \dfrac{12.60 - 13.00}{0.22\ \sqrt{(8 + 8)/64}}$ (6-5)

$t = 3.64$

3. Look up for degrees of freedom = ∞, in Table 6.2.

3. The value t = 3.64 is greater than any value in the DF = ∞ line, in table 6.1; therefore, the probability that the difference is due to chance alone is less than .001 for a two-tailed test.

4. If t (computed in step 2 ≤ tdf ∞ (value from table in step 3), decide that X_1 and X_2 differ with regard to their average levels, then-

Conclude that the difference in lead content is significant.(4) Conclude that the difference in lead content is significant.

---

[3] Adapted by permission from *Quality Control Handbook*, edited by J. M. Juran, 2nd Ed. Copyright 1962 by McGraw-Hill, Inc. Used with permission of McGraw-Hill Book Co.

[4] Each sample should be subjected to a x2 test (see Test 5, following on page 56).

4. The "t" Test for Difference Between Two Sample Means ($\overline{X}_1$ and $\overline{X}_2$)[5]
This form of the "t" test is applied when the standard deviation σ' is unknown, but believed to be the same for both populations. In this example, we will assume that sets of ten identical samples of toluene on charcoal were analyzed by two different laboratories, with the following results:

	Lab A	Lab B
No. of samples analyzed	10	10
Avg. mg/l = $\overline{X}$ mg/l	1.609	1.530
Variability σ mg/l	0.0844	0.1104

The question to be answered, then, is: Can it be stated confidently that the two laboratories' analyses results are significantly different? The value 1.609 - 1.503 = .079 is the difference under consideration. Since this test assumes equal variances, they should be tested first, using the procedure in Test 6, following on page.

Procedure	Example

1. Compute $\sigma' = \sqrt{\dfrac{n_1 \sigma_1^2 + n_2 \sigma_2^2}{n_1 + n_2 - 2}}$

1. $\sigma' \sqrt{\dfrac{10(.0844)^2 + 10(.1104)^2}{10 + 10 - 2}}$

$$(6\text{-}6)$$

$$= 0.1035$$

2. Compute $t = \sigma' \sqrt{\dfrac{n_1 + n_2}{n_1 n_2}}$ numerator $X_1 - X_2$

2. $t = \dfrac{1.609 - 1.530}{.1035 \sqrt{\dfrac{10 + 10}{100}}}$

$$(6\text{-}7)$$

0.0463

$$= 0.079 = 1.706$$

3. Compute DF = $n_1 + n_2$ - 2.
4. Look up the value of t for DF = 18, in Table 6.2, at a preselected level of significance. (in this case, 5%).
5. Compare the value of t in step 4, with the calculated value of t in step 2. If the calculated t value is larger than the table value, decide that X_1 and X_2 differ, with regards to their average concentration levels, at the selected level of significance.

3. DF = 10 + 10 -2 = 18.
4. t_{18} (a = .05) = 2.101

5. Since t = 1.706 is less than t = 2.101 (from the table), decide that the means are not significantly differently, at the a = .05 level of significance.

[5]Adapted by permission from Quality Control Handbook, edited by J. M. Juran, 2nd Ed. Copyright 1962 by McGraw-Hill, Inc. Used with permission of McGraw-Hill Book Co.

5. The Chi-Square (χ^2) Test for the Difference Between the Sample Variability (σ) and the Population Variability (σ')[6]

When using this test, the population variability is a known quantity. A decrease in the uniformity of test samples coming from the same source may be as important as a shift in the average value. If the test results have been close, such a loss of precision may have the effect of generating a serious number of questionable analyses. On the other hand, detecting increased uniformity and the reason for it may pave the way to a permanent improvement in quality.

As an example of the application of the Chi-square test, let us use the data tabulated for the urine specimens in Test 1. The sample standard deviation, when computed, is found to be 0.078 σg Hg/liter. We know, from past experience, that the population standard deviation is 0.25, μg Hg/liter, as given in the Test 1 example. The question posed here is: Does the low value of 0.078 indicate that the new sample is significantly more uniform?

Procedure	Example

1. Compute:

$$\chi^2 = n \left(\frac{\sigma}{\sigma'} \right)$$

1. $\chi^2 = 12 \left(\dfrac{.078}{0.25} \right) 2$

    $= 1.168$ (6-8)

2. Look up the value nearest 1.168 for n - 1 = 11 degrees of freedom in the "Distribution of χ^2" table (Table 6.3).

2. Since 1.168 falls to the left of all those tabulated, we conclude that the probability is well above .99.

3. For χ^2, a very low P (below 0.01) is interpreted as meaning that the sample σ is significantly small.

3. We conclude that the sample standard deviation is significantly small. Values in the new sample were significantly more uniform in mercury content than in previous samples.

Note: Values of χ^2 are given for DF, up to 120. For larger samples, use can be made of the fact that the distribution of t is approaching the normal curve. A test with:

$$t = \frac{(s - \sigma) \sqrt{2n}}{\sigma}$$

(6-9)

may be made, finding the probabilities in the table of t values (Table 6.1), on line DF = ∞.

---

[6]Adapted by permission from *Quality Control Handbook*, edited by J. M. Juran, 2nd Ed. Copyright 1962 by McGraw-Hill, Inc. Used with permission of McGraw-Hill Book Co.

Table 6.3. * Distribution of χ^2

Values of χ^2 corresponding to certain selected probabilities (i.e., tail areas under the curve). To illustrate; the probability is 0.95 that a sample with 20 degrees of freedom, taken from a normal distribution, would have $\chi^2 = 31.41$ or smaller.

VALUES OF $x^2{}_p$ CORRESPONDING TO P

DF	$\chi^2.005$	$\chi^2.01$	$\chi^2.025$	$\chi^2.05$	$\chi^2.10$	$\chi^2.90$	$\chi^2.95$	$\chi^2.975$	$\chi^2.99$	$\chi^2.995$
1	0.000039	0.00016	0.00098	0.0039	0.0158	2.71	3.84	5.02	6.63	7.88
2	0.0100	0.0201	0.0506	0.10226	0.2107	4.61	5.99	7.38	9.21	10.60
3	0.0717	0.115	0.16	0.352	0.584	6.25	7.81	9.35	11.34	12.84
4	0.207	0.297	0.484	0.711	1.064	7.78	9.49	11.14	13.28	14.86
5	0.412	0.554	0.831	1.15	1.61	9.24	11.07	12.83	15.09	16.75
6	0.676	0.872	1.24	1.64	2.20	10.64	12.59	14.45	16.81	18.55
7	0.989	1.24	1.69	2.17	2.83	12.02	14.07	16.01	18.48	20.28
8	1.34	1.65	2.18	2.73	3.49	13.36	15.51	17.53	20.09	21.96
9	1.73	2.09	2.70	3.33	4.17	14.68	16.92	19.02	211.67	23.59
10	2.16	2.56	3.25	3.94	4.87	15.99	18.31	20.48	23.21	25.19
11	2.60	3.05	3.82	4.57	5.558	17.28	19.68	21.92	24.73	26.76
12	3.07	3.57	4.40	5.23	6.30	18.55	21.03	23.34	26.22	28.30
13	3.57	4.11	5.01	5.89	7.04	19.81	22.36	24.74	27.69	29.82
14	4.07	4.66	5.63	6.57	7.779	21.06	23.68	26.12	29.14	31.32
15	4.60	5.23	6.26	7.26	8.55	22.31	25.00	27.49	30.58	32.80
16	5.14	5.81	6.91	7.96	9.31	23.54	26.30	28.85	32.00	34.27
18	6.26	7.01	8.23	9.39	10.86	25.99	28.87	3.153	34.81	37.16
20	7.43	8.26	9.59	10.85	12.44	28.41	31.41	34.17	37.57	40.00
24	9.89	10.86	12.40	13.85	15.66	33.20	36.42	39.36	4.298	4.556
30	13.76	14.95	16.79	18.59	20.60	40.26	43.77	46.98	50.89	53.67
40	20.71	22.16	24.43	26.51	29.05	51.81	55.76	59.34	63.69	66.77
60	35.53	37.48	40.48	43.19	46.46	74.40	79.08	83.30	88.38	9.195
120	83.85	86.92	91.58	95.70	100.62	140.23	146.57	152.21	158.95	163.64

5a.   Alternate Method of Testing for Difference Between Sample Variability and Population Variability

Presented here is another way to test for the significance of the difference between the standard deviations of the population and the sample.

This is a simpler approach than that of the previous test, but again, the population standard deviation must be known. For this test, we will use Table 6.4, "Significant Ratios for s/ σ = G Normal Population." The question to be answered here is the same as that of the preceding test: Is the new sample significantly more uniform?

Procedure	Example
1. Compute: $G = \dfrac{s}{\sigma}$	1. $G = \dfrac{s}{\sigma} = \dfrac{.078}{.25}$ $G = .312$ (6-10)
2. Look up G for N = 12 in Table 6.3.	2. For N = 12, the lowest table value .504 is greater than .312.
3. Compare G value (.312) and the table value in step 2. If computed G (.312) is less than the table value, conclude that a significant difference exists.	3. We conclude that there is a significant difference.

6.   The "F" Test for Difference in Variability (σ a^2 and σb^2) In Two Samples[7]

In this case, we will use the data from Test 4, and determine whether the laboratory results differ in variability; in other words, are the results of Laboratory A more consistent than those of Laboratory B?

Procedure	Example
1. Compute: $F = \dfrac{\sigma\, b^2}{\sigma\, a^2}$	1. $F = \dfrac{.1104^2}{.0844^2} = 1.71$ (6-11)
2. Note degrees of freedom.	2. $DF_1 - n_1 - 1 = 9$ $DF_2 - n_2 - 1 = 9$
3. Enter Table 6.5 (Distribution of F) at column 9, row 9, to obtain F.9 (9.9) **IMPORTANT!** Note that the DF line heading the table requires that the greater mean square amount be placed in the numerator.	3. F.9 (9,9) = 3.18
4. Compare step 1 with step 3. If computed F is greater than the table is greater than the table value in step 3, the variabilities are significantly different.	4. 1.71 < 3.18 Decide that the data does not disclose a significant difference between laboratories, and, if a real difference exists, the present samples are too small to disclose it.

---

[7]Adapted by permission from Quality Control Handbook, edited by J. M. Juran, 2nd E. Copyright 1962 by McGraw-Hill, Inc. Used with permission of McGraw-Hill Book Co.

Table 6.4   Significant Ratios for σ/.σ' = G Normal Population

No. of Cases, N, in Sample	Probability of as Great or a Greater Value of σ/σ'				
	.99	.95	.05	.01	.001
2	.009	.044	1.386	1.821	2.327
3	.082	.185+	1.413	1.752	2.146
4	.170	.297	1.398	1.684	2.107
5	.244	.377	1.378	1.630	1.922
6	.304	.437	1.358	1.586	1.849
7	.353	.483	1.341	1.550	1.791
8	.394	.520	1.326	1.520	1.744
9	.428	.551	1.313	1.494	1.704
10	.457	.577	1.301	1.472	1.670
11	.482	.598	1.290	1.453	1.640
12	.504	.617	1.280	1.435+	1.614
13	.524	.634	1.272	1.420	1.591
14	.542	.649	1.264	1.406	1.570
15	.557	.662	1.257	1.394	1.552
16	.572	.674	1.250	1.382	1.535-
17	.585-	.684	1.244	1.372	1.520
18	.597	.694	1.238	1.362	1.505+
19	.608	.703	1.233	1.353	1.492
20	.618	.711	1.228	1.345+	1.480
21	.627	.719	1.223	1.337	1.469
22	.636	.726	1.219	1.330	1.458
23	.644	.732	1.214	1.324	1.449
24	.652	.739	1.211	1.317	1.439
25	.659	.744	1.207	1.311	1.431
26	.666	.750	1.203	1.306	1.423
27	.672	.755-	1.200	1.300	1.414
28	.678	.759	1.197	1.295+	1.408
29	.684	.764	1.194	1.290	1.401
30	.689	.768	1.191	1.286	1.394
31	.695-	.772	1.188	1.281	1.388
36	.713	.787	1.178	1.266	1.365+
40	.732	.801	1.168	1.249	1.342
45	.748	.814	1.159	1.236	1.323
50	.761	.824	1.152	1.224	1.307
55	.772	.832	1.145+	1.214	1.292
60	.782	.840	1.140	1.205+	1.280
65	.791	.847	1.135-	1.198	1.269
70	.799	.853	1.130	1.191	1.260
75	.806	.858	1.126	1.184	1.251
80	.812	.863	1.122	1.179	1.243
85	.818	.867	1.119	1.174	1.236
90	.823	.871	1.116	1.169	1.229
95	.828	.874	1.113	1.164	1.223
100	.832	.878	1.110	1.160	1.218
200	.882	.915-	1.079	1.114	1.154
300	.904	.931	1.065+	1.094	1.126
400	.917	.940	1.57	1.061	1.109
500	.926	.947	1.051	1.073	1.098
1000	.948	.963	1.036	1.052	1.069

Table 6.5   Distribution of F*

5 Per Cent (Roman Type) and 1 Per Cent (Boldface Type)

Values of F corresponding to two selected probabilities (*i.e.*, tail areas under the curve). To illustrate: the probability is 0.05 that the ratio of two mean squares obtained with 20 and 10 degrees of freedom in numerator and denominator, respectively, would yield $F = 2.77$ or larger.

DF_1 degrees of freedom for greater mean square (placed in the numerator)

DF_2	1	2	3	4	5	6	7	8	9	10	11	12	14	16	20	24	30	40	50	75	100	200	500	∞
1	161	200	216	225	230	234	237	239	241	242	243	244	245	246	248	249	250	251	252	253	253	254	254	254
	4,052	**4,999**	**5,403**	**5,625**	**5,764**	**5,859**	**5,928**	**5,981**	**6,022**	**6,056**	**6,082**	**6,106**	**6,142**	**6,169**	**6,208**	**6,234**	**6,258**	**6,286**	**6,302**	**6,323**	**6,334**	**6,352**	**6,361**	**6,366**
2	18.51	19.00	19.16	19.25	19.30	19.33	19.36	19.37	19.38	19.39	19.40	19.41	19.42	19.43	19.44	19.45	19.46	19.47	19.47	19.48	19.49	19.49	19.50	19.50
	98.49	**99.00**	**99.17**	**99.25**	**99.30**	**99.33**	**99.34**	**99.36**	**99.38**	**99.40**	**99.41**	**99.42**	**99.43**	**99.44**	**99.45**	**99.46**	**99.47**	**99.48**	**99.48**	**99.49**	**99.49**	**99.49**	**99.50**	**99.50**
3	10.13	9.55	9.28	9.12	9.01	8.94	8.88	8.84	8.81	8.78	8.76	8.74	8.71	8.69	8.66	8.64	8.62	8.60	8.58	8.57	8.56	8.54	8.54	8.53
	34.12	**30.82**	**29.46**	**28.71**	**28.24**	**27.91**	**27.67**	**27.49**	**27.34**	**27.23**	**27.13**	**27.05**	**26.92**	**26.83**	**26.69**	**26.60**	**26.50**	**26.41**	**26.36**	**26.27**	**26.23**	**26.18**	**26.14**	**26.12**
4	7.71	6.94	6.59	6.39	6.26	6.16	6.09	6.04	6.00	5.96	5.93	5.91	5.87	5.84	5.80	5.77	5.74	5.71	5.70	5.68	5.66	5.65	5.64	5.63
	21.20	**18.00**	**16.69**	**15.98**	**15.52**	**15.21**	**14.98**	**14.80**	**14.66**	**14.54**	**14.45**	**14.37**	**14.24**	**14.15**	**14.02**	**13.93**	**13.83**	**13.74**	**13.69**	**13.61**	**13.57**	**13.52**	**13.48**	**13.46**
5	6.61	5.79	5.41	5.19	5.05	4.95	4.88	4.82	4.78	4.74	4.70	4.68	4.64	4.60	4.56	4.53	4.50	4.46	4.44	4.42	4.40	4.38	4.37	4.36
	16.26	**13.27**	**12.06**	**11.39**	**10.97**	**10.67**	**10.45**	**10.27**	**10.15**	**10.05**	**9.96**	**9.89**	**9.77**	**9.68**	**9.55**	**9.47**	**9.38**	**9.29**	**9.24**	**9.17**	**9.13**	**9.07**	**9.04**	**9.02**
6	5.99	5.14	4.76	4.53	4.39	4.28	4.21	4.15	4.10	4.06	4.03	4.00	3.96	3.92	3.87	3.84	3.81	3.77	3.75	3.72	3.71	3.69	3.68	3.67
	13.74	**10.92**	**9.78**	**9.15**	**8.75**	**8.47**	**8.26**	**8.10**	**7.98**	**7.87**	**7.79**	**7.72**	**7.60**	**7.52**	**7.39**	**7.31**	**7.23**	**7.14**	**7.09**	**7.02**	**6.99**	**6.94**	**6.90**	**6.88**
7	5.59	4.74	4.35	4.12	3.97	3.87	3.79	3.73	3.68	3.63	3.60	3.57	3.52	3.49	3.44	3.41	3.38	3.34	3.32	3.29	3.28	3.25	3.24	3.23
	12.25	**9.55**	**8.45**	**7.85**	**7.46**	**7.19**	**7.00**	**6.84**	**6.71**	**6.62**	**6.54**	**6.47**	**6.35**	**6.27**	**6.15**	**6.07**	**5.98**	**5.90**	**5.85**	**5.78**	**5.75**	**5.70**	**5.67**	**5.65**
8	5.32	4.46	4.07	3.84	3.69	3.58	3.50	3.44	3.39	3.34	3.31	3.28	3.23	3.20	3.15	3.12	3.08	3.05	3.03	3.00	2.98	2.96	2.94	2.93
	11.26	**8.65**	**7.59**	**7.01**	**6.63**	**6.37**	**6.19**	**6.03**	**5.91**	**5.82**	**5.74**	**5.67**	**5.56**	**5.48**	**5.36**	**5.28**	**5.20**	**5.11**	**5.06**	**5.00**	**4.96**	**4.91**	**4.88**	**4.86**
9	5.12	4.26	3.86	3.63	3.48	3.37	3.29	3.23	3.18	3.13	3.10	3.07	3.02	2.98	2.93	2.90	2.86	2.82	2.80	2.77	2.76	2.73	2.72	2.71
	10.56	**8.02**	**6.99**	**6.42**	**6.06**	**5.80**	**5.62**	**5.47**	**5.35**	**5.26**	**5.18**	**5.11**	**5.00**	**4.92**	**4.80**	**4.73**	**4.64**	**4.56**	**4.51**	**4.45**	**4.41**	**4.36**	**4.33**	**4.31**
10	4.96	4.10	3.71	3.48	3.33	3.22	3.14	3.07	3.02	2.97	2.94	2.91	2.86	2.82	2.77	2.74	2.70	2.67	2.64	2.61	2.59	2.56	2.55	2.54
	10.04	**7.56**	**6.55**	**5.99**	**5.64**	**5.39**	**5.21**	**5.06**	**4.95**	**4.85**	**4.78**	**4.71**	**4.60**	**4.52**	**4.41**	**4.33**	**4.25**	**4.17**	**4.12**	**4.05**	**4.01**	**3.96**	**3.93**	**3.91**

Reprinted by permission from *Statistical Methods*, 8th Edition by G. Snedecor and W. Cochran, © 1989 by Iowa State University Press, Ames.

Table 6.5   (Continued)

DF₁ degrees of freedom for greater mean square (placed in the numerator)

DF₂	1	2	3	4	5	6	7	8	9	10	11	12	14	16	20	24	30	40	50	75	100	200	500	∞
11	4.84	3.98	3.59	3.36	3.20	3.09	3.01	2.95	2.90	2.86	2.82	2.79	2.74	2.70	2.65	2.61	2.57	2.53	2.50	2.47	2.45	2.42	2.41	2.40
	9.65	**7.20**	**6.22**	**5.67**	**5.32**	**5.07**	**4.88**	**4.74**	**4.63**	**4.54**	**4.46**	**4.40**	**4.29**	**4.21**	**4.10**	**4.02**	**3.94**	**3.86**	**3.80**	**3.74**	**3.70**	**3.66**	**3.62**	**3.60**
12	4.75	3.88	3.49	3.26	3.11	3.00	2.92	2.85	2.80	2.76	2.72	2.69	2.64	2.60	2.54	2.50	2.46	2.42	2.40	2.36	2.35	2.32	2.31	2.30
	9.33	**6.93**	**5.95**	**5.41**	**5.06**	**4.82**	**4.65**	**4.50**	**4.39**	**4.30**	**4.22**	**4.16**	**4.05**	**3.98**	**3.86**	**3.78**	**3.70**	**3.61**	**3.56**	**3.49**	**3.46**	**3.41**	**3.38**	**3.36**
13	4.67	3.80	3.41	3.18	3.02	2.92	2.84	2.77	2.72	2.67	2.63	2.60	2.55	2.51	2.46	2.42	2.38	2.34	2.32	2.28	2.26	2.24	2.22	2.21
	9.07	**6.70**	**5.74**	**5.20**	**4.86**	**4.62**	**4.44**	**4.30**	**4.19**	**4.10**	**4.02**	**3.96**	**3.85**	**3.78**	**3.67**	**3.59**	**3.51**	**3.42**	**3.37**	**3.30**	**3.27**	**3.21**	**3.18**	**3.16**
14	4.60	3.74	3.34	3.11	2.96	2.85	2.77	2.70	2.65	2.60	2.56	2.53	2.48	2.44	2.39	2.35	2.31	2.27	2.24	2.21	2.19	2.16	2.14	2.13
	8.86	**6.51**	**5.56**	**5.03**	**4.69**	**4.46**	**4.28**	**4.14**	**4.03**	**3.94**	**3.86**	**3.80**	**3.70**	**3.62**	**3.51**	**3.43**	**3.34**	**3.26**	**3.21**	**3.14**	**3.11**	**3.06**	**3.02**	**3.00**
15	4.54	3.68	3.29	3.06	2.90	2.79	2.70	2.64	2.59	2.55	2.51	2.48	2.43	2.39	2.33	2.29	2.25	2.21	2.18	2.15	2.12	2.10	2.08	2.07
	8.68	**6.36**	**5.42**	**4.89**	**4.56**	**4.32**	**4.14**	**4.00**	**3.89**	**3.80**	**3.73**	**3.67**	**3.56**	**3.48**	**3.36**	**3.29**	**3.20**	**3.12**	**3.07**	**3.00**	**2.97**	**2.92**	**2.89**	**2.87**
16	4.49	3.63	3.24	3.01	2.85	2.74	2.66	2.59	2.54	2.49	2.45	2.42	2.37	2.33	2.28	2.24	2.20	2.16	2.13	2.09	2.07	2.04	2.02	2.01
	8.53	**6.23**	**5.29**	**4.77**	**4.44**	**4.20**	**4.03**	**3.89**	**3.78**	**3.69**	**3.61**	**3.55**	**3.45**	**3.37**	**3.25**	**3.18**	**3.10**	**3.01**	**2.96**	**2.89**	**2.86**	**2.80**	**2.77**	**2.75**
17	4.45	3.59	3.20	2.96	2.81	2.70	2.62	2.55	2.50	2.45	2.41	2.38	2.33	2.29	2.23	2.19	2.15	2.11	2.08	2.04	2.02	1.99	1.97	1.96
	8.40	**6.11**	**5.18**	**4.67**	**4.34**	**4.10**	**3.93**	**3.79**	**3.68**	**3.59**	**3.52**	**3.45**	**3.35**	**3.27**	**3.16**	**3.08**	**3.00**	**2.92**	**2.86**	**2.79**	**2.76**	**2.70**	**2.67**	**2.65**
18	4.41	3.55	3.16	2.93	2.77	2.66	2.58	2.51	2.46	2.41	2.37	2.34	2.29	2.25	2.19	2.15	2.11	2.07	2.04	2.00	1.98	1.95	1.93	1.92
	8.28	**6.01**	**5.09**	**4.58**	**4.25**	**4.01**	**3.85**	**3.71**	**3.60**	**3.51**	**3.44**	**3.37**	**3.27**	**3.19**	**3.07**	**3.00**	**2.91**	**2.83**	**2.78**	**2.71**	**2.68**	**2.62**	**2.59**	**2.57**
19	4.38	3.52	3.13	2.90	2.74	2.63	2.55	2.48	2.43	2.38	2.34	2.31	2.26	2.21	2.15	2.11	2.07	2.02	2.00	1.96	1.94	1.91	1.90	1.88
	8.18	**5.93**	**5.01**	**4.50**	**4.17**	**3.94**	**3.77**	**3.63**	**3.52**	**3.43**	**3.36**	**3.30**	**3.19**	**3.12**	**3.00**	**2.92**	**2.84**	**2.76**	**2.70**	**2.63**	**2.60**	**2.54**	**2.51**	**2.49**
20	4.35	3.49	3.10	2.87	2.71	2.60	2.52	2.45	2.40	2.35	2.31	2.28	2.23	2.18	2.12	2.08	2.04	1.99	1.96	1.92	1.90	1.87	1.85	1.84
	8.10	**5.85**	**4.94**	**4.43**	**4.10**	**3.87**	**3.71**	**3.56**	**3.45**	**3.37**	**3.30**	**3.23**	**3.13**	**3.05**	**2.94**	**2.86**	**2.77**	**2.69**	**2.63**	**2.56**	**2.53**	**2.47**	**2.44**	**2.42**
21	4.32	3.47	3.07	2.84	2.68	2.57	2.49	2.42	2.37	2.32	2.28	2.25	2.20	2.15	2.09	2.05	2.00	1.96	1.93	1.89	1.87	1.84	1.82	1.81
	8.02	**5.78**	**4.87**	**4.37**	**4.04**	**3.81**	**3.65**	**3.51**	**3.40**	**3.31**	**3.24**	**3.17**	**3.07**	**2.99**	**2.88**	**2.80**	**2.72**	**2.63**	**2.58**	**2.51**	**2.47**	**2.42**	**2.38**	**2.36**
22	4.30	3.44	3.05	2.82	2.66	2.55	2.47	2.40	2.35	2.30	2.26	2.23	2.18	2.13	2.07	2.03	1.98	1.93	1.91	1.87	1.84	1.81	1.80	1.78
	7.94	**5.72**	**4.82**	**4.31**	**3.99**	**3.76**	**3.59**	**3.45**	**3.35**	**3.26**	**3.18**	**3.12**	**3.02**	**2.94**	**2.83**	**2.75**	**2.67**	**2.58**	**2.53**	**2.46**	**2.42**	**2.37**	**2.33**	**2.31**
23	4.28	3.42	3.03	2.80	2.64	2.53	2.45	2.38	2.32	2.28	2.24	2.20	2.14	2.10	2.04	2.00	1.96	1.91	1.88	1.84	1.82	1.79	1.77	1.76

Table 6.5 (Continued)

DF_1 degrees of freedom for greater mean square (placed in the numerator)

DF_2	1	2	3	4	5	6	7	8	9	10	11	12	14	16	20	24	30	40	50	75	100	200	500	∞
	7.88	5.66	4.76	4.26	3.94	3.71	3.54	3.41	3.30	3.21	3.14	3.07	2.97	2.89	2.78	2.70	2.62	2.53	2.48	2.41	2.37	2.32	2.28	2.26
24	4.26	3.40	3.01	2.78	2.62	2.51	2.43	2.36	2.30	2.26	2.22	2.18	2.13	2.09	2.02	1.98	1.94	1.89	1.86	1.82	1.80	1.76	1.74	1.73
	7.82	5.61	4.72	4.22	3.90	3.67	3.50	3.36	3.25	3.17	3.09	3.03	2.93	2.85	2.74	2.66	2.58	2.49	2.44	2.36	2.33	2.27	2.23	2.21
25	4.24	3.38	2.99	2.76	2.60	2.49	2.41	2.34	2.28	2.24	2.20	2.16	2.11	2.06	2.00	1.96	1.92	1.87	1.84	1.80	1.77	1.74	1.72	1.71
	7.77	5.57	4.68	4.18	3.86	3.63	3.46	3.32	3.21	3.13	3.05	2.99	2.89	2.81	2.70	2.62	2.54	2.45	2.40	2.32	2.29	2.23	2.19	2.17
26	4.22	3.37	2.98	2.74	2.59	2.47	2.39	2.32	2.27	2.22	2.18	2.15	2.10	2.05	1.99	1.95	1.90	1.85	1.82	1.78	1.76	1.72	1.70	1.69
	7.72	5.53	4.64	4.14	3.82	3.59	3.42	3.29	3.17	3.09	3.02	2.96	2.86	2.77	2.66	2.58	2.50	2.41	2.36	2.28	2.25	2.19	2.15	2.13
27	4.21	3.35	2.96	2.73	2.57	2.46	2.37	2.30	2.25	2.20	2.16	2.13	2.08	2.03	1.97	1.93	1.88	1.84	1.80	1.76	1.74	1.71	1.68	1.67
	7.68	5.49	4.60	4.11	3.79	3.56	3.39	3.26	3.14	3.06	2.98	2.93	2.83	2.74	2.63	2.55	2.47	2.38	2.33	2.25	2.21	2.16	2.12	2.10
28	4.20	3.34	2.95	2.71	2.56	2.44	2.36	2.29	2.24	2.19	2.15	2.12	2.06	2.02	1.96	1.91	1.87	1.81	1.78	1.75	1.72	1.69	1.67	1.65
	7.64	5.45	4.57	4.07	3.76	3.53	3.36	3.23	3.11	3.03	2.95	2.90	2.80	2.71	2.60	2.52	2.44	2.35	2.30	2.22	2.18	2.13	2.09	2.06
29	4.18	3.33	2.93	2.70	2.55	2.43	2.35	2.28	2.22	2.18	2.14	2.10	2.05	2.00	1.94	1.90	1.85	1.80	1.77	1.73	1.71	1.68	1.65	1.64
	7.60	5.42	4.54	4.04	3.73	3.50	3.33	3.20	3.08	3.00	2.92	2.87	2.77	2.68	2.57	2.49	2.41	2.32	2.27	2.19	2.15	2.10	2.06	2.03
30	4.17	3.32	2.92	2.69	2.53	2.42	2.34	2.27	2.21	2.16	2.12	2.09	2.04	1.99	1.93	1.89	1.84	1.79	1.76	1.72	1.69	1.66	1.64	1.62
	7.56	5.39	4.51	4.02	3.70	3.47	3.30	3.17	3.06	2.98	2.90	2.84	2.74	2.66	2.55	2.47	2.38	2.29	2.24	2.16	2.13	2.07	2.03	2.01
32	4.15	3.30	2.90	2.67	2.51	2.40	2.32	2.25	2.19	2.14	2.10	2.07	2.02	1.97	1.91	1.86	1.82	1.76	1.74	1.69	1.67	1.64	1.61	1.59
	7.50	5.34	4.46	3.97	3.66	3.42	3.25	3.12	3.01	2.94	2.86	2.80	2.70	2.62	2.51	2.42	2.34	2.25	2.20	2.12	2.08	2.02	1.98	1.96
34	4.13	3.28	2.88	2.65	2.49	2.38	2.30	2.23	2.17	2.12	2.08	2.05	2.00	1.95	1.89	1.84	1.80	1.74	1.71	1.67	1.64	1.61	1.59	1.57
	7.44	5.29	4.42	3.93	3.61	3.38	3.21	3.08	2.97	2.89	2.82	2.76	2.66	2.58	2.47	2.38	2.30	2.21	2.16	2.08	2.04	1.98	1.94	1.91
36	4.11	3.26	2.86	2.63	2.48	2.36	2.28	2.21	2.15	2.10	2.06	2.03	1.98	1.93	1.87	1.82	1.78	1.72	1.69	1.65	1.62	1.59	1.56	1.55
	7.39	5.25	4.38	3.89	3.58	3.35	3.18	3.04	2.94	2.86	2.78	2.72	2.62	2.54	2.43	2.35	2.26	2.17	2.12	2.04	2.00	1.94	1.90	1.87
38	4.10	3.25	2.85	2.62	2.46	2.35	2.26	2.19	2.14	2.09	2.05	2.02	1.96	1.92	1.85	1.80	1.76	1.71	1.67	1.63	1.60	1.57	1.54	1.53
	7.35	5.21	4.34	3.86	3.54	3.32	3.15	3.02	2.91	2.82	2.75	2.69	2.59	2.51	2.40	2.32	2.22	2.14	2.08	2.00	1.97	1.90	1.86	1.84
40	4.08	3.23	2.84	2.61	2.45	2.34	2.25	2.18	2.12	2.07	2.04	2.00	1.95	1.90	1.84	1.79	1.74	1.69	1.66	1.61	1.59	1.55	1.53	1.51
	7.31	5.18	4.31	3.83	3.51	3.29	3.12	2.99	2.88	2.80	2.73	2.66	2.56	2.49	2.37	2.29	2.20	2.11	2.05	1.97	1.94	1.88	1.84	1.81

Table 6.5   (Continued)

DF$_1$ degrees of freedom for greater mean square (placed in the numerator)

DF$_2$	1	2	3	4	5	6	7	8	9	10	11	12	14	16	20	24	30	40	50	75	100	200	500	∞
42	4.07	3.22	2.83	2.59	2.44	2.32	2.24	2.17	2.11	2.06	2.02	1.99	1.94	1.89	1.82	1.78	1.73	1.68	1.64	1.60	1.57	1.54	1.51	1.49
	7.27	5.15	4.29	3.80	3.49	3.26	3.10	2.96	2.86	2.77	2.70	2.64	2.54	2.46	2.35	2.26	2.17	2.08	2.02	1.94	1.91	1.85	1.80	1.78
44	4.06	3.21	2.82	2.58	2.43	2.31	2.23	2.16	2.10	2.05	2.01	1.98	1.92	1.88	1.81	1.76	1.72	1.66	1.63	1.58	1.56	1.52	1.50	1.48
	7.24	5.12	4.26	3.78	3.46	3.24	3.07	2.94	2.84	2.75	2.68	2.62	2.52	2.44	2.32	2.24	2.15	2.06	2.00	1.92	1.88	1.82	1.78	1.75
46	4.05	3.20	2.81	2.57	2.42	2.30	2.22	2.14	2.09	2.04	2.00	1.97	1.91	1.87	1.80	1.75	1.71	1.65	1.62	1.57	1.54	1.51	1.48	1.46
	7.21	5.10	4.24	3.76	3.44	3.22	3.05	2.92	2.82	2.73	2.66	2.60	2.50	2.42	2.30	2.22	2.13	2.04	1.98	1.90	1.86	1.80	1.76	1.72
48	4.04	3.19	2.80	2.56	2.41	2.30	2.21	2.14	2.08	2.03	1.99	1.96	1.90	1.86	1.79	1.74	1.70	1.64	1.61	1.56	1.53	1.50	1.47	1.45
	7.19	5.08	4.22	3.74	3.42	3.20	3.04	2.90	2.80	2.71	2.64	2.58	2.48	2.40	2.28	2.20	2.11	2.02	1.96	1.88	1.84	1.78	1.73	1.70
50	4.03	3.18	2.79	2.56	2.40	2.29	2.20	2.13	2.07	2.02	1.98	1.95	1.90	1.85	1.78	1.74	1.69	1.63	1.60	1.55	1.52	1.48	1.46	1.44
	7.17	5.06	4.20	3.72	3.41	3.18	3.02	2.88	2.78	2.70	2.62	2.56	2.46	2.39	2.26	2.18	2.10	2.00	1.94	1.86	1.82	1.76	1.71	1.68
55	4.02	3.17	2.78	2.54	2.38	2.27	2.18	2.11	2.05	2.00	1.97	1.93	1.88	1.83	1.76	1.72	1.67	1.61	1.58	1.52	1.50	1.46	1.43	1.41
	7.12	5.01	4.16	3.68	3.37	3.15	2.98	2.85	2.75	2.66	2.59	2.53	2.43	2.35	2.23	2.15	2.06	1.96	1.90	1.82	1.78	1.71	1.66	1.64
60	4.00	3.15	2.76	2.52	2.37	2.25	2.17	2.10	2.04	1.99	1.95	1.92	1.86	1.81	1.75	1.70	1.65	1.59	1.56	1.50	1.48	1.44	1.41	1.39
	7.08	4.98	4.13	3.65	3.34	3.12	2.95	2.82	2.72	2.63	2.56	2.50	2.40	2.32	2.20	2.12	2.03	1.93	1.87	1.79	1.74	1.68	1.63	1.60
65	3.99	3.14	2.75	2.51	2.36	2.24	2.15	2.08	2.02	1.98	1.94	1.90	1.85	1.80	1.73	1.68	1.63	1.57	1.54	1.49	1.46	1.42	1.39	1.37
	7.04	4.95	4.10	3.62	3.31	3.09	2.93	2.79	2.70	2.61	2.54	2.47	2.37	2.30	2.18	2.09	2.00	1.90	1.84	1.76	1.71	1.64	1.60	1.56
70	3.98	3.13	2.74	2.50	2.35	2.23	2.14	2.07	2.01	1.97	1.93	1.89	1.84	1.79	1.72	1.67	1.62	1.56	1.53	1.47	1.45	1.40	1.37	1.35
	7.01	4.92	4.08	3.60	3.29	3.07	2.91	2.77	2.67	2.59	2.51	2.45	2.35	2.28	2.15	2.07	1.98	1.88	1.82	1.74	1.69	1.62	1.56	1.53
80	3.96	3.11	2.72	2.48	2.33	2.21	2.12	2.05	1.99	1.95	1.91	1.88	1.82	1.77	1.70	1.65	1.60	1.54	1.51	1.45	1.42	1.38	1.35	1.32
	6.96	4.88	4.04	3.56	3.25	3.04	2.87	2.74	2.64	2.55	2.48	2.41	2.32	2.24	2.11	2.03	1.94	1.84	1.78	1.70	165	1.57	1.52	1.49
100	3.94	3.09	2.70	2.46	2.30	2.19	2.10	2.03	1.97	1.92	1.88	1.85	1.79	1.75	1.68	1.63	1.57	1.51	1.48	1.42	1.39	1.34	1.30	1.28
	6.90	4.82	3.98	3.51	3.20	2.99	2.82	2.69	2.59	2.51	2.43	2.36	2.26	2.19	2.06	1.98	1.89	1.79	1.73	1.64	1.59	1.51	1.46	1.43
125	3.92	3.07	2.68	2.44	2.29	2.17	2.08	2.01	1.95	1.90	1.86	1.83	1.77	1.72	1.65	1.60	1.55	1.49	1.45	1.39	1.36	1.31	1.27	1.25
	6.84	4.78	3.94	3.47	3.17	2.95	2.79	2.65	2.56	2.47	2.40	2.33	2.23	2.15	2.03	1.94	1.85	1.75	1.68	1.59	1.54	1.46	1.40	1.37
150	3.91	3.06	2.67	2.43	2.27	2.16	2.07	2.00	1.94	1.89	1.85	1.82	1.76	1.71	1.64	1.59	1.54	1.47	1.44	1.37	1.34	1.29	1.25	1.22

Table 6.5 *(Continued)*

DF$_1$ degrees of freedom for greater mean square (placed in the numerator)

DF$_2$	1	2	3	4	5	6	7	8	9	10	11	12	14	16	20	24	30	40	50	75	100	200	500	∞
200	**6.81**	**4.75**	**3.91**	**3.44**	**3.14**	**2.92**	**2.76**	**2.62**	**2.53**	**2.44**	**2.37**	**2.30**	**2.20**	**2.12**	**2.00**	**1.91**	**1.83**	**1.72**	**1.66**	**1.56**	**1.51**	**1.43**	**1.37**	**1.33**
	3.89	3.04	2.65	2.41	2.26	2.14	2.05	1.98	1.92	1.87	1.83	1.80	1.74	1.69	1.62	1.57	1.52	1.45	1.42	1.35	1.32	1.26	1.22	1.19
400	**6.76**	**4.71**	**3.88**	**3.41**	**3.11**	**2.90**	**2.73**	**2.60**	**2.50**	**2.41**	**2.34**	**2.28**	**2.17**	**2.09**	**1.97**	**1.88**	**1.79**	**1.69**	**1.62**	**1.53**	**1.48**	**1.39**	**1.33**	**1.28**
	3.86	3.02	2.62	2.39	2.23	2.12	2.03	1.96	1.90	1.85	1.81	1.78	1.72	1.67	1.60	1.54	1.49	1.42	1.38	1.32	1.28	1.22	1.16	1.13
1000	**6.70**	**4.66**	**3.83**	**3.36**	**3.06**	**2.85**	**2.69**	**2.55**	**2.46**	**2.37**	**2.29**	**2.23**	**2.12**	**2.04**	**1.92**	**1.84**	**1.74**	**1.64**	**1.57**	**1.47**	**1.42**	**1.32**	**1.24**	**1.19**
	3.85	3.00	2.61	2.38	2.22	2.10	2.02	1.95	1.89	1.84	1.80	1.76	1.70	1.65	1.58	1.53	1.47	1.41	1.36	1.30	1.26	1.19	1.13	1.08
	6.66	**4.62**	**3.80**	**3.34**	**3.04**	**2.82**	**2.66**	**2.53**	**2.43**	**2.34**	**2.26**	**2.20**	**2.09**	**2.01**	**1.89**	**1.81**	**1.71**	**1.61**	**1.54**	**1.44**	**1.38**	**1.28**	**1.19**	**1.11**
∞	3.84	2.99	2.60	2.37	2.21	2.09	2.01	1.94	1.88	1.83	1.79	1.75	1.69	1.64	1.57	1.52	1.46	1.40	1.35	1.28	1.24	1.17	1.11	1.00
	6.64	**4.60**	**3.78**	**3.32**	**3.02**	**2.80**	**2.64**	**2.51**	**2.41**	**2.32**	**2.24**	**2.18**	**2.07**	**1.99**	**1.87**	**1.79**	**1.69**	**1.59**	**1.52**	**1.41**	**1.36**	**1.25**	**1.15**	**1.00**

Two additional statistical tools are presented here that are often useful in interpreting laboratory data. The first of these is a variation of the use of the Range R to estimate the standard deviation. We have already encountered this, when discussing control charts (Chapter 4).

An acceptable approximation of sigma is obtained by dividing the average Range R by a factor, d_2 taken from Table 6.6, "Factors For Estimating the Standard Deviation From the Range."

Assume that we have ten samples of two each, of analyses of toluene on charcoal, with the results shown below:

Sample Pair No.	Range R Between Values in Pairs (mg/L)
1	0.18
2	0.04
3	0.20
4	0.14
5	0.13
6	0.03
7	0.11
8	0.18
9	0.07
10	0.13
Total =	1.21
Average Range =	0.121

Table 6.6   Factors for Estimating the Standard Deviation from the Range

Size of Sample	$\dfrac{1}{d_2}$	d_2
2	.8865	1.128
3	.5907	1.693
4	.4857	2.059
5	.4299	2.326
6	.3946	2.543
7	.3698	2.709
8	.3512	2.847
9	.3367	2.970
10	.3249	3.077
12	.3069	3.258
16	.2831	3.532

Procedure	Example
1. Compute the average $\overline{R}$.	1. $\dfrac{1.21}{10} = .121$
2. Find d_2 for sample size 2 in Table 6.6.	2. $d_2 = 1.128$
3. Compute estimate of: $\hat{\sigma} = \dfrac{\overline{R}}{d_2}$	3. $\hat{\sigma} = \dfrac{.121}{1.128} = .1073$

$$(6-12)$$

Another useful technique for interpreting data generated in laboratory work is the Analysis of Variance, usually known by its acronym, ANOVA. Remember that the variance, σ^2, is merely the standard deviation squared, which is another way of describing the variability of a set of data. Analyzing data, in order to determine if there is a significant amount of variability when more than two sets of data are involved, is an important tool in our inventory of statistical techniques. A short cut to ANOVA is provided by John Tukey's "Gap and Straggler Test," which will be described here.

Consider the results of analyses for mercury in urine, which were reported by three different laboratories that tested aliquots of the same samples, as follows:

Sample No.	Lab A	Lab B	Lab C
1	.126	.100	.09
2	.062	.048	.04
3	.050	.033	.03
4	.060	.046	.05
5	.102	.090	.06
6	.158	.178	.11
7	.046	.038	.03
8	.038	.045	.03
9	.042	.043	.02
10	.034	.033	.03
$\Sigma =$	.718	.653	.490
$R =$	.124	.145	.090

The question to be answered is: Is there a significant difference in laboratory performance at 1% level of significance?

Table 6.7 Critical Factors For One-way (balanced) Divisions Into Groups*
The four entries, are respectively, for:

Range with 5% risk   Gap with 5% risk
Range with 1% risk   Gap with 1% risk

and are to be multiplied by the sum of the ranges within groups. (Risks are on per experiment basis.)

Number in group = number per range

Each cell shows four entries arranged as: (Range 5% risk, Gap 5% risk) on the top line and (Range 1% risk, Gap 1% risk) on the bottom line.

Number of groups = number of summed	Risk	2 Range	2 Gap	3 Range	3 Gap	4 Range	4 Gap	5 Range	5 Gap	6 Range	6 Gap	7 Range	7 Gap	8 Range	8 Gap	9 Range	9 Gap	10 Range	10 Gap
2	5%	3.43	3.43	1.91	1.91	1.63	1.63	1.53	1.53	1.50	1.50	1.49	1.49	1.49	1.49	1.50	1.50	1.52	1.52
	1%	7.92	7.92	3.14	3.14	2.47	2.47	2.24	2.24	2.14	2.14	2.10	2.10	2.08	2.08	2.09	2.09	2.09	2.09
3	5%	2.37	1.76	1.44	1.14	1.25	1.02	1.19	0.98	1.18	0.96	1.17	0.96	1.17	0.97	1.18	0.98	1.20	0.99
	1%	4.42	3.25	2.14	1.73	1.74	1.47	1.60	1.37	1.55	1.32	1.53	1.33	1.52	1.33	1.53	1.34	1.55	1.35
4	5%	1.98	1.18	1.13	0.81	1.01	0.74	0.94	0.72	0.92	0.71	0.92	0.71	0.94	0.72	0.96	0.73	0.97	0.73
	1%	2.96	1.96	1.57	1.19	1.33	1.04	1.24	0.98	1.21	0.96	1.21	0.96	1.21	0.97	1.22	0.98	1.23	0.99
5	5%	1.40	0.88	0.94	0.63	0.85	0.58	0.81	0.56	0.80	0.56	0.80	0.56	0.81	0.57	0.82	0.58	0.84	0.59
	1%	2.06	1.39	1.25	0.91	1.08	0.80	1.02	0.77	0.99	0.76	0.99	0.76	0.99	0.77	1.00	0.77	1.01	0.78
6	5%	1.16	0.70	0.81	0.52	0.75	0.48	0.69	0.47	0.69	0.46	0.69	0.47	0.70	0.47	0.71	0.48	0.72	0.49
	1%	1.69	1.07	1.04	0.73	0.94	0.66	0.86	0.63	0.85	0.62	0.85	0.63	0.85	0.63	0.85	0.64	0.86	0.65
7	5%	1.00	0.58	0.70	0.44	0.63	0.40	0.61	0.40	0.61	0.40	0.61	0.40	0.62	0.41	0.63	0.41	0.63	0.42
	1%	1.39	0.87	0.89	0.61	0.78	0.55	0.75	0.54	0.74	0.53	0.74	0.53	0.74	0.54	0.75	0.55	0.76	0.55
8	5%	0.87	0.50	0.62	0.38	0.57	0.35	0.55	0.34	0.55	0.34	0.55	0.35	0.55	0.35	0.56	0.36	0.57	0.37
	1%	1.20	0.74	0.78	0.53	0.69	0.48	0.66	0.47	0.65	0.46	0.65	0.46	0.66	0.47	0.66	0.48	0.67	0.48
9	5%	0.78	0.44	0.56	0.33	0.51	0.31	0.50	0.30	0.49	0.30	0.50	0.31	0.50	0.31	0.51	0.31	0.52	0.32
	1%	1.03	0.63	0.71	0.46	0.62	0.44	0.62	0.44	0.59	0.43	0.59	0.42	0.59	0.42	0.60	0.43	0.61	0.43
10	5%	0.70	0.39	0.51	0.30	0.46	0.28	0.45	0.27	0.45	0.27	0.45	0.28	0.46	0.28	0.47	0.28	0.47	0.29
	1%	0.91	0.56	0.62	0.41	0.57	0.38	0.54	0.37	0.53	0.36	0.54	0.37	0.54	0.37	0.55	0.38	0.55	0.38

*Reproduced with permission from Table 1 of John W. Tukey, "Quick and Dirty Methods in Statistics. II. Simple Analysis for Standard Design," Quality Control. Conference Papers, Fifth Annual Convention, American Society, American Society for Quality Control. May 23 and 14, 1951, p. 194.

Procedure	Example

1. Compute the sum and range of each column of data.	1. *Lab ABC*
	    Column Sum.718.653.490
	    Column R.124.145.090

2. Compute the range of sums.	2. .718 - .490 = .228
3. Compute the sum of ranges.	3. .124 + .145 + .090 = .359
4. Multiply the sum of ranges by a critical value found in Table 6.7 (1.55).[8]	4. 1.55 (0.359) = 0.5564
5. If the range of sums in step 2 is greater than the product found in step 4, there is a significant difference associated with the critical value. Otherwise, the differences are not significant.	5. Since the rage of sums (0.228) is less than 0.5564, we cannot state that a significant difference exists between laboratory analytical results.

References

Juran, J. M. 1962. *Quality Control Handbook.* 2d Ed. New York: McGraw-Hill Book Company.

Natrella, Mary G. 1963. *Experimental Statistics.* Washington, D.C.: National Bureau of Standards.

---

[8]The critical value, 1.55, is chosen from the box shown in column 10 (ten samples), third row (three laboratories). Since, in this example, we are using the range in our calculations, we choose 1.55, in the lower left-hand corner of the box, since it represents the critical value for the range, at the 1% risk level. (Note the table heading carefully.)

7

Sampling Plans

Laboratories will often need to use a sampling plan, in order to select a sample of data points or data sheets from the voluminous amount of information generated during testing or analytical work. In addition to this, some of the larger laboratories that buy testing or analytical materials, such as air sampling filters and gas detector tubes in large quantities, may wish to inspect lots of such goods received, in order to make sure that the products meet specifications. It should be pointed out here that 100% inspection of lots of materials is not considered to be economically practical. Not only that, but, because of inspector error caused by fatigue, inattention, boredom, distractions, and so on, 100% inspection is not 100% effective. Indeed, it should be assumed that screening every item in a lot will be only about 80% effective. In other words, the inspector will miss two defective or nonconforming items out of every ten present in the lot being screened.

Normally, instead of resorting to 100% inspection, sampling plans are used in conducting such inspections. Sampling plans afford a high degree of assurance to the user that acceptable materials are passed into the laboratory for use.

MIL-STD-105D, "Sampling Procedures and Tables for Inspection by Attributes," and its equivalent in the private sector, ANSI/ASQC Standard Z1.4-1981, are the most commonly used sampling plans. They, or adapted and translated versions of them, are used throughout the world. These attribute sampling plans are known as "producer's risk" schemes. That is, the producer or supplier of an item bears the risk of having an acceptable or "good" lot rejected, when inspected under the provisions of the plan. For the two plans mentioned above, the confidence level for the plan results is set at about 95%. This confidence level is built into the tables published in the standards.

This chapter will discuss, in some detail, using the single sampling plans presented in MIL-STD-105D and ANSI/ASQC Z1.4. These standards also include double and multiple plans that work in very much the same way as single sampling, so that an understanding of how single sampling works will form a basis for using double and multiple sampling, too. We will also briefly discuss some other types of sampling plans and will look at two kinds of sampling plans that should not be used and discuss why.

The efficiency of any sampling plan is described by its operating characteristic curve (OC curve). The OC curve is characterized by "N," the lot size, "n" the sample size, "c" the acceptance number, and "p" the process average. OC curves generally take the shape of a reversed "S." The ideal OC curve is Z-shaped, as in Figure 7.1. This idealistic example of an OC curve would discriminate perfectly against product that has a process average of 3% defective or worse and will accept product that has a process average of less than 3%. On the OC curve, the process average, p, is read along the bottom scale (abscissa), and the probability of acceptance, P_a, is read on the vertical scale (ordinate). OC curves are constucted from Poisson Tables (Summation of Terms of the Exponential Limit) or, more easily, by using a Thorndike Chart (Figure 7.2). In the example illustrated, an arbitrarily selected sample size of 100 is used, together with process average values along the bottom scale, and the specified acceptance figure (in this case, 3). To use the chart, one multiplies the selected p's by the lot size, 100, which yields, in this example, the values np = 2.0 and 5.0. The chart is entered along the bottom line at values 2.0 and 5.0 and traced vertically until the acceptance value curve (c = 3) is met. At the junctures of the c = 3 curve and the 2.0 and 5.0 vertical traces, horizontal lines are constructed to the left to meet the ordinate scale, which shows the Probability of Observing c or Less Defects. The values obtained at the juncture of the horizontal traces and the ordinate are then transferred to the OC Curve chart, immediately to the left of the main chart. The observed values become the P_a values, which are plotted at the juncture of the corresponding np values. The example shows two such plots and the resulting data points. Plotting four or five such data points should be sufficient in most cases, to allow the plot of the OC Curve, for the particular sampling plan to be described.

There are other terms associated with attribute sampling plans that are of importance to the user. The first is the "Acceptable Quality Level," or "AQL." We will quote, from MIL-STD-105D, the definition of an AQL and then explain what it means and what it does not mean. Thus, an AQL is:

the maximum percent defective or the maximum number of defective parts per hundred units that, for the purposes of sampling inspection, can be considered satisfactory as a process average.

When an AQL is specified at a certain level (say 1%), this does not mean that 1% of all the pieces from lots inspected and accepted will be defective. For instance, if there is an accepted lot of 10,000 pieces, it does not mean that 100

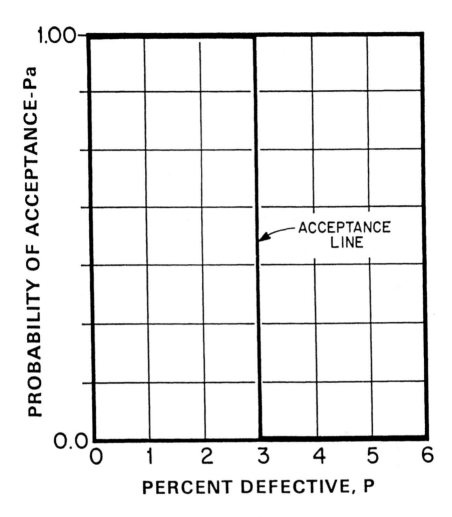

FIGURE 7.1.   Copied by permission from *Quality Control Handbook*, edited by J. M. Juran, 4th Ed. Copyright 1988 by McGraw-Hill Book Company. Used with permission of McGraw-Hill Book Company.

pieces will be found to be defective. What it does mean is that, if a large number of lots are inspected by using a specified AQL of 1% the user runs the risk of accepting 1% of the lots, thinking that they are good, when they actually should not have been accepted. Using the sampling plan does not offer any information about the quality of individual lots, because the AQL alone does not describe the protection to the consumer for individual lots or batches, but, more directly, relates to what may be expected from a series of lots or batches, provided that the steps

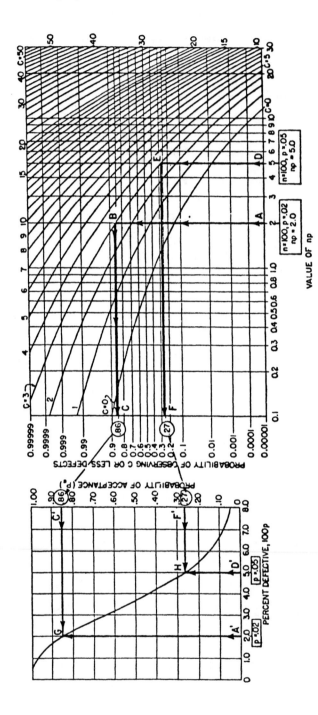

FIGURE 7.2.   Procedure for obtaining the operating characteristic curve. Sample (n) = 100, allowable number of defects in the sample (c) = 3. Copied by permission from *Quality Control Handbook*, edited by J. M. Juran, 3rd Ed. Copyright 1974 by McGraw-Hill Book Company. Used with permission of McGraw-Hill Book Company.

involving switching procedures indicated in Par. 8.3 of MIL-STD-105D are followed.

Another term associated with sampling plans is the "Average Outgoing Quality," or "AOQ." The AOQ is the average quality of outgoing product, including all accepted and rejected lots or batches, after the rejected lots or batches have been effectively 100% inspected and all the defectives replaced by non-defectives. The third term found is the "Average Outgoing Quality Limit," or "AOQL." The AOQL is the maximum of the AOQs, for all possible incoming quantities for a given acceptance plan. Looking at the AOQ curve shown in Figure 7.3 shows that the AOQL is the highest point on the AOQ curve. Note that the slanting line indicates what that OQ curve would look like, without 100% inspection of rejected lots. We calculate the AOQ curve by establishing representative points of the computation $AOQ = p(P_a)$, for various product qualities at various probabilities of acceptance. This particular sampling plan is drawn from an OC curved described by n = 75 and c = 1. We would like to stress again here the importance of taking random samples from lots being inspected, to determine whether they are acceptable and to repeat that the selection of the sample should be randomized by by the use of dice, playing cards, tables of random numbers, or a calculator with a random number generator

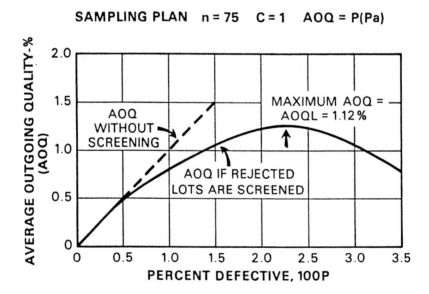

FIGURE 7.3.   Copied by permission from *Quality Control Handbook*, edited by J. M. Juran, 4th Ed. Copyright 1988 McGraw-Hill Book Company. Used with permission of McGraw-Hill Company.

function. Again, every precaution should be taken, to eliminate bias from the sample chosen.

In the past, it has been common practice for those not familiar with using sampling plans to use such simplified schemes as, "We will always take 10% of the lot as our sample." Unfortunately, this is not an effective way to take samples, as illustrated by Figure 7.4. This chart shows four OC curves drawn for lot sizes

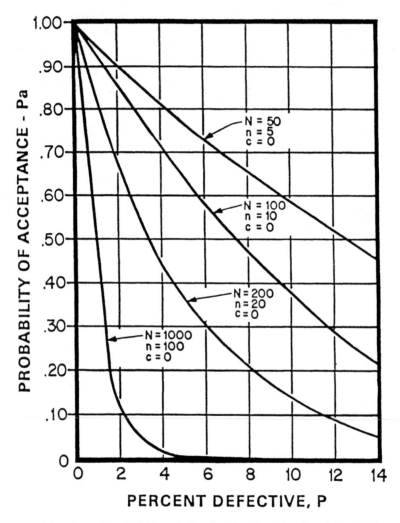

FIGURE 7.4.   Comparison of OC Curves for Four Sampling Plans Using a Sample Size of 10% of the lLot Size. Copied by permission form Statistical Quality Control, by Grant & Leavenworth, 4th Ed. Copyright 1972 by McGraw-Hill Book Company. Used with permission of McGraw-Hill Book Company.

50, 100, 200, and 1000. Note that for the smaller lots, such as N = 50, using a sample of n = 5 results in a 75% chance of accepting lots that are 6% defective. For N = 100, we run a 57% chance of accepting lots that are 6% defective. It is not until we have larger lots that the sample provides adequate protection against the acceptance of bad lots. This is illustrated by the case of N = 1000, where a sample size n = 100, which indicates a 10% chance of accepting a lot that is only 2% defective. Thus, the reader will see from this chart that the use of a fixed percentage of the lot as a sample is not an effective way to conduct sampling.

Another sampling plan that is easy to establish, but still not effective, is to take a fixed sample size, regardless of the lot size. This case is illustrated by Figure 7.5, where a sample of 20 units is used for all lot sizes. Here we find that, for all three of the given lot sizes (n = 1000, n = 200, and n = 100), we have about a 30% chance of accepting lots that are 6% defective, which is not good. It is not until we have a small lot of n = 50 that the sampling plan begins to work for us, because the sample equals about 40% of the lot, which is a very large sample for a lot or batch of this size. In analyzing the effectiveness of these two sampling plans, we have seen how the OC curve describes the degree of protection afforded by a given sampling plan.

Since the attribute sampling plans, MIL-Std-105D and ANSI/ASQC Z-1.4, are the most widely used plans in the world, we will describe here how to use the single sampling plans set forth in these standards. A series of tables extracted from MIL-STD-105D will be used, to familiarize the reader with the use of these tables.

Before proceeding, it is necessary to explain the source of AQLs specified to be used in conjunction with MIL-STD-105D.

Typically, AQLs are assigned to dimensions and other call-outs appearing on drawings as a matter of engineering judgment, based on tolerances, criticality of possible failures, design requirements, eventual use, and location of the dimension or call-out, with regard to mating parts. AQLs appear in a document commonly called the "Inspection Instruction/Classification of Defects" sheet. This form habitually accompanies each engineering drawing during inspection and serves as the inspector's guidance as to how he or she will inspect the item in question. An example appears in Table 7.1. Note that the user receives, in addition to the item identification, the following information:

1. The sampling plan to be used;
2. What aspect on the drawing is to be inspected;
3. What instrument or gage is to be used for the inspection; and
4. What AQL is to be used as a basis for judgment as to its acceptability.

A form such as this is provided for each part being inspected, with AQLs being assigned for Critical, Major "A," Major "B," and Minor defects that may appear

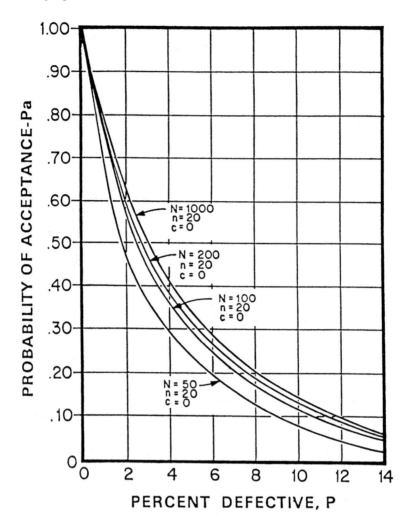

FIGURE 7.5.   Comparison of OC Curves for Four Sampling Plans Using the Same Sample Size (20), Each With c = 0. Copied by permission from Statistical Quality Control, by Grand & Leavenworth, 4th Ed. Copyright 1972 by McGraw-Hill Book Company. Used with permission of McGraw-Hill Book Company.

in the parts being inspected. Definitions of these classifications are quoted from Par. 2.1 of MIL-STD-105, as follows:

Note: ANSI/ASQC Z-1. 4 does not provide for the classification of defects, but
    since this practice illustrates the source of AQL specifications, it is included here.
Critical defect. A critical defect is a defect that judgment and experience indi-
    cate is likely to result in hazardous or unsafe conditions for individuals using,

Table 7.1   Classification of Defects Inspection Instructions

(2) <u>Part No.</u>: 000000	(4) <u>Description</u>: Face Piece
(3) <u>Drawing No.</u>: F000000	(3) <u>Specification</u>: None
(4) <u>Drawing Title</u>: Rubber Gadget	
(5) <u>Inspection Location</u>: In-process	(6) <u>Lot Size</u>: 8 hr. shift

(7) <u>Sample Procedure</u>: MIL-STD-105D, Level II unless noted otherwise.

(8) <u>Remarks</u>: Item numbers refer to numbers marked on drawings or specification.

(9) <u>Critical</u>

(10) Inspect 100%. No defects allowed.

(12) 1 1.50″ ± .01″ inside dia. (2 holes)	Go-No/Go Plug Gage(11)
2 Leakage	XYZ 100 procedure

<u>Major "A"</u>
AQL 1.0%

101 Check lens for cracks, scratches, chips, etc. (Note 3 on drawing)	Visual
102 Shore Durometer (70 ± 5)A	Shore Durometer
103 .125″ ± .010″ thickness	Vernier Calipers

<u>Major "B"</u>
AQL 2.5%

201 Check rubber for imperfections (Note 2 on drawing)	Visual
202 6″ x 4″ ± 1/8″	Scale

<u>Minor</u>
AQL 4.0%

302 Appearance (color)	Visual
303 All other dimensions	Under 4″ Calipers 4″ & Over Scale

maintaining, or depending upon the product, or a defect that judgment and experience indicate is likely to prevent performance of the tactical function of a major end item, such as a ship, aircraft, tank, missile, or space vehicle.

Major defect.[1] A major defect is a defect, other than critical, that is likely to result in failure, or to materially reduce the usability of the unit of product for its intended purpose.

Minor defect.[2] A minor defect is a defect that is not likely to materially reduce the usability of the product, for its intended purpose, or is a departure from estab-

---

[1]Defects may be, by choice, grouped into other classes, or into subclasses within these classes, as is the case in Major "A" and Major "B" (Table 7.1)

[2]Minor defects are usually, but not always, defects that are cosmetic in nature, affecting the appearance of the item. Examples are variations in color, scratches, mars, blemishes, incorrect markings, and so forth.

lished standards, having little bearing on the effective use or operation of the unit.

The next question faced is how to employ the specified AQLs, when using sampling tables from MIL-STD-105D. For the purposes of illustration, we will assume that we have received a lot of 1000 pieces of some product presented for inspection. The first task is to determine the sample size to be use. Go first to Table I—Sample size code letters (Table 7-2). Go to the column headed "General inspection levels." Using inspection level II[3], find the sample size code letter "J" for lot size 501 to 1000. Now turn to Table IIA—Single sampling plans for normal inspection (Master table) (Figure 7-6) and find, from the two left-hand columns that the sample size for Sample size code letter J is 80. The next step is to find the acceptance and rejection criteria values. Assuming that the item to be inspected has a specified AQL of 1%, go to the column headed 1.0. Go down the 1.0 column to the J-80 row and find that the acceptance number is 1, and the rejection number is 2. This means that if, during the inspection of the sample of 80 pieces, 1 defective part is found, that part is discarded and the lot is accepted; however, if 2 or more detectives are found, the lot is rejected and set aside for disposition.

While our example suggests AQLs of 1.0%, 2.5%, and 4.0%, these were selected for the purposes of illustration only. Current practices, dictated by marketing pressures, particularly in the automotive and electronics industries suggest that

Table 7.2   Sample size code letters   TABLE I

Lot or batch size			Special inspection levels				General inspection levels		
			S-1	S-2	S-3	S-4	I	II	III
2	to	8	A	A	A	A	A	A	B
9	to	15	A	A	A	A	A	B	C
16	to	25	A	A	B	B	B	C	D
26	to	50	A	B	B	C	C	D	E
51	to	90	B	B	C	C	C	E	F
91	to	150	B	B	C	D	D	F	G
151	to	280	B	C	D	E	E	G	H
281	to	500	B	C	D	E	F	H	J
501	to	1200	C	C	E	F	G	J	K
1201	to	3200	C	D	E	G	H	K	L
3201	to	10000	C	D	F	G	J	L	M
10001	to	35000	C	D	F	H	K	M	N
35001	to	150000	D	E	G	J	K	N	P
150001	to	500000	D	E	G	J	L	N	P
500001	and	over	D	E	H	K	N	Q	R

[3]Par. 9.2 MIL-STD-105D stipulates: "Unless otherwise specified, Inspection Level II will be used."

SINGLE
NORMAL

TABLE II A

Sample size code letter	Sample size	0.010	0.015	0.025	0.040	0.065	0.10	0.15	0.25	0.40	0.65	1.0	1.5	2.5	4.0	6.5	10	15	25	40	65	100	150	250	400	650	1000
		Acceptable Quality Levels (normal inspection)																									
		Ac Re	Ac Re	Ac Re	Ac Re	Ac Re	Ac Re	Ac Re	Ac Re	Ac Re	Ac Re	Ac Re	Ac Re	Ac Re	Ac Re	Ac Re	Ac Re	Ac Re	Ac Re	Ac Re	Ac Re	Ac Re	Ac Re	Ac Re	Ac Re	Ac Re	Ac Re
A	2	↓	↓	↓	↓	↓	↓	↓	↓	↓	↓	↓	↓	↓	↓	0 1	↑	↓	1 2	2 3	3 4	5 6	7 8	10 11	14 15	21 22	30 31
B	3	↓	↓	↓	↓	↓	↓	↓	↓	↓	↓	↓	↓	↓	0 1	↑	↓	1 2	2 3	3 4	5 6	7 8	10 11	14 15	21 22	30 31	44 45
C	5	↓	↓	↓	↓	↓	↓	↓	↓	↓	↓	↓	↓	0 1	↑	↓	1 2	2 3	3 4	5 6	7 8	10 11	14 15	21 22	30 31	44 45	↑
D	8	↓	↓	↓	↓	↓	↓	↓	↓	↓	↓	↓	0 1	↑	↓	1 2	2 3	3 4	5 6	7 8	10 11	14 15	21 22	30 31	44 45	↑	↑
E	13	↓	↓	↓	↓	↓	↓	↓	↓	↓	↓	0 1	↑	↓	1 2	2 3	3 4	5 6	7 8	10 11	14 15	21 22	30 31	44 45	↑	↑	↑
F	20	↓	↓	↓	↓	↓	↓	↓	↓	↓	0 1	↑	↓	1 2	2 3	3 4	5 6	7 8	10 11	14 15	21 22	↑	↑	↑	↑	↑	↑
G	32	↓	↓	↓	↓	↓	↓	↓	↓	0 1	↑	↓	1 2	2 3	3 4	5 6	7 8	10 11	14 15	21 22	↑	↑	↑	↑	↑	↑	↑
H	50	↓	↓	↓	↓	↓	↓	↓	0 1	↑	↓	1 2	2 3	3 4	5 6	7 8	10 11	14 15	21 22	↑	↑	↑	↑	↑	↑	↑	↑
J	80	↓	↓	↓	↓	↓	↓	0 1	↑	↓	1 2	2 3	3 4	5 6	7 8	10 11	14 15	21 22	↑	↑	↑	↑	↑	↑	↑	↑	↑
K	125	↓	↓	↓	↓	↓	0 1	↑	↓	1 2	2 3	3 4	5 6	7 8	10 11	14 15	21 22	↑	↑	↑	↑	↑	↑	↑	↑	↑	↑
L	200	↓	↓	↓	↓	0 1	↑	↓	1 2	2 3	3 4	5 6	7 8	10 11	14 15	21 22	↑	↑	↑	↑	↑	↑	↑	↑	↑	↑	↑
M	315	↓	↓	↓	0 1	↑	↓	1 2	2 3	3 4	5 6	7 8	10 11	14 15	21 22	↑	↑	↑	↑	↑	↑	↑	↑	↑	↑	↑	↑
N	500	↓	↓	0 1	↑	↓	1 2	2 3	3 4	5 6	7 8	10 11	14 15	21 22	↑	↑	↑	↑	↑	↑	↑	↑	↑	↑	↑	↑	↑
P	800	↓	0 1	↑	↓	1 2	2 3	3 4	5 6	7 8	10 11	14 15	21 22	↑	↑	↑	↑	↑	↑	↑	↑	↑	↑	↑	↑	↑	↑
Q	1250	0 1	↑	↓	1 2	2 3	3 4	5 6	7 8	10 11	14 15	21 22	↑	↑	↑	↑	↑	↑	↑	↑	↑	↑	↑	↑	↑	↑	↑
R	2000	↑	↑	1 2	2 3	3 4	5 6	7 8	10 11	14 15	21 22	↑	↑	↑	↑	↑	↑	↑	↑	↑	↑	↑	↑	↑	↑	↑	↑

↓ = Use first sampling plan below arrow. If sample size equals, or exceeds, lot or batch size, do 100 percent inspection.
↑ = Use first sampling plan above arrow.
Ac = Acceptance number.
Re = Rejection number.

FIGURE 7.6 Single Sampling Plans for Normal Inspection (Master Table). = Use first sampling plan below. If sample size equals, or exceeds, lot or batch size, do 100 percent inspection. = Use first sampling plan above arrow. Ac = Acceptance number. Re = Rejection number.

the use of AQLs below the 0.010% (100 parts per million) level will be common practice in the years to come.

This sampling system rewards good quality, by reducing the required sample size in its provision for a switching procedure to be followed in certain cases. If ten consecutive lots of a product are accepted, the sampling plan shown in Table II-C—Single sampling plans for reduced inspection (Master table) (Figure 7.7) may be used. Table II-C shows that a sample of only 32, less than half the previous sample of 80, may now be used for a lot size of 1000. Going across to the 1.0% AQL column, we see that the acceptance number is now 1 and the rejection number is 3. The question immediately arises- What if there are two defects? Following the instructions at the dagger symbol at the bottom of the chart, we find that: "If the acceptance number has been exceeded, but the rejection number has not been reached, accept the lot, but reinstate normal inspection." A finding of two defects in this case is known as the "area of indecision."

In the same manner, findings of "bad" quality are penalized by the requirement to switch to tightened inspection. If two lots out of five, when at normal inspection, are rejected, the switching procedure again takes effect. This time it requires going to Table II-B—Single sampling plans for tightened inspection (Master table) (Figure 7.8). Again, enter the table at Sample size code letter J, and find that the sample size has not increased, but remains at 80. However, going across to the AQL = 1.0% column, we see that the acceptance and rejection numbers have changed to 1 and 2 respectively, instead of the 2 and 3 that we had at normal inspection. Thus, the sampling scheme has been tightened, by squeezing down the acceptance and rejection levels. One may return to normal inspection from tightened inspection when five consecutive lots or batches have been found to be acceptable.

Examination of Figure 7.9, which is a comparison of the OC curves for the sampling plan for sample size code K at 1% AQL, shows the degree of protection offered at each level of sampling.

MIL-STD-105D is by no means the only sampling plan available. Its counterparts, MIL-STD-414, "Sampling Procedures and Tables For Inspection by Variables For Percent Defective," and ANSI/ASQC Z-1.9, "Sampling Procedures and Tables for Inspection by Variables for Percent Nonconforming," require fewer samples to be taken and are, therefore, appropriate for use, where testing or inspection methods are expensive or destructive. Their use, however, requires the acquiring results in terms of values, not "good or bad" or "go-no-go," and thus requires more record keeping. Calculations are necessary, in order to arrive at standard deviations that are needed to enter the tables. These tables (Figure 7.10) are similar to those of MIL-STD-105D and are used in much the same way. The decision to use MIL-STD-414 must be based on balancing the desire to take fewer samples against the cost of increased paperwork and computation time.

There are many other kinds of sampling plans, some of which are suitable only

SINGLE
REDUCED

TABLE II C

Acceptable Quality Levels (reduced inspection)†

Sample size code letter	Sample size	0.010	0.015	0.025	0.040	0.065	0.10	0.15	0.25	0.40	0.65
		Ac Re	Ac Re	Ac Re	Ac Re	Ac Re	Ac Re	Ac Re	Ac Re	Ac Re	Ac Re
A	2	↓	↓	↓	↓	↓	↓	↓	↓	↓	↓
B	2	↓	↓	↓	↓	↓	↓	↓	↓	↓	↓
C	2	↓	↓	↓	↓	↓	↓	↓	↓	↓	↓
D	3	↓	↓	↓	↓	↓	↓	↓	↓	↓	↓
E	5	↓	↓	↓	↓	↓	↓	↓	↓	↓	↓
F	8	↓	↓	↓	↓	↓	↓	↓	↓	↓	0 1
G	13	↓	↓	↓	↓	↓	↓	↓	↓	0 1	↑
H	20	↓	↓	↓	↓	↓	↓	↓	0 1	↑	↓
J	32	↓	↓	↓	↓	↓	↓	0 1	↑	↓	0 2
K	50	↓	↓	↓	↓	↓	0 1	↑	↓	0 2	1 3
L	80	↓	↓	↓	↓	0 1	↑	↓	0 2	1 3	1 4
M	125	↓	↓	↓	0 1	↑	↓	0 2	1 3	1 4	2 5
N	200	↓	↓	0 1	↑	↓	0 2	1 3	1 4	2 5	3 6
P	315	↓	0 1	↑	↓	0 2	1 3	1 4	2 5	3 6	5 8
Q	500	0 1	↑	↓	0 2	1 3	1 4	2 5	3 6	5 8	7 10
R	800	↑	↑	0 2	1 3	1 4	2 5	3 6	5 8	7 10	10 13

Sample size code letter	Sample size	1.0	1.5	2.5	4.0	6.5	10	15	25	40	65	100	150	250	400	650	1000
		Ac Re	Ac Re	Ac Re	Ac Re	Ac Re	Ac Re	Ac Re	Ac Re	Ac Re	Ac Re	Ac Re	Ac Re	Ac Re	Ac Re	Ac Re	Ac Re
A	2	↓	↓	↓	↓	0 1	↓	↓	1 2	2 3	3 4	5 6	7 8	10 11	14 15	21 22	30 31
B	2	↓	↓	↓	0 1	↑	↓	0 2	1 3	2 4	3 5	5 6	7 8	10 11	14 15	21 22	30 31
C	2	↓	↓	0 1	↑	↓	0 2	1 3	1 4	2 5	3 6	5 8	7 10	10 13	14 17	21 24	↑
D	3	↓	0 1	↑	↓	0 2	1 3	1 4	2 5	3 6	5 8	7 10	10 13	14 17	21 24	↑	↑
E	5	0 1	↑	↓	0 2	1 3	1 4	2 5	3 6	5 8	7 10	10 13	14 17	21 24	↑	↑	↑
F	8	↑	↓	0 2	1 3	1 4	2 5	3 6	5 8	7 10	10 13	↑	↑	↑	↑	↑	↑
G	13	↓	0 2	1 3	1 4	2 5	3 6	5 8	7 10	10 13	↑	↑	↑	↑	↑	↑	↑
H	20	0 2	1 3	1 4	2 5	3 6	5 8	7 10	10 13	↑	↑	↑	↑	↑	↑	↑	↑
J	32	1 3	1 4	2 5	3 6	5 8	7 10	10 13	↑	↑	↑	↑	↑	↑	↑	↑	↑
K	50	1 4	2 5	3 6	5 8	7 10	10 13	↑	↑	↑	↑	↑	↑	↑	↑	↑	↑
L	80	2 5	3 6	5 8	7 10	10 13	↑	↑	↑	↑	↑	↑	↑	↑	↑	↑	↑
M	125	3 6	5 8	7 10	10 13	↑	↑	↑	↑	↑	↑	↑	↑	↑	↑	↑	↑
N	200	5 8	7 10	10 13	↑	↑	↑	↑	↑	↑	↑	↑	↑	↑	↑	↑	↑
P	315	7 10	10 13	↑	↑	↑	↑	↑	↑	↑	↑	↑	↑	↑	↑	↑	↑
Q	500	10 13	↑	↑	↑	↑	↑	↑	↑	↑	↑	↑	↑	↑	↑	↑	↑
R	800	↑	↑	↑	↑	↑	↑	↑	↑	↑	↑	↑	↑	↑	↑	↑	↑

↓ = Use first sampling plan below arrow. If sample size equals or exceeds lot or batch size, do 100 percent inspection.
↑ = Use first sampling plan above arrow.
Ac = Acceptance number.
Re = Rejection number.
† = If the acceptance number has been exceeded, but the rejection number has not been reached, accept the lot, but reinstate normal inspection (see 10.1.4).

FIGURE 7.7 Single Sampling Plans for Reduced Inspection (Master Table). = Use first sampling plan below. If sample size equals, or exceeds, lot or batch size, do 100 percent inspection. = Use first sampling plan above arrow. Ac = Acceptance number. Re = Rejection number.

TABLE II B

Sample size code letter	Sample size	Acceptable Quality Levels (tightened inspection)																									
		0.010	0.015	0.025	0.040	0.065	0.10	0.15	0.25	0.40	0.65	1.0	1.5	2.5	4.0	6.5	10	15	25	40	65	100	150	250	400	650	1000
		Ac Re	Ac Re	Ac Re	Ac Re	Ac Re	Ac Re	Ac Re	Ac Re	Ac Re	Ac Re	Ac Re	Ac Re	Ac Re	Ac Re	Ac Re	Ac Re	Ac Re	Ac Re	Ac Re	Ac Re	Ac Re	Ac Re	Ac Re	Ac Re	Ac Re	Ac Re
A	2	↓	↓	↓	↓	↓	↓	↓	↓	↓	↓	↓	↓	↓	↓	↓	↓	↓	↓	1 2	2 3	3 4	5 6	8 9	12 13	18 19	27 28
B	3	↓	↓	↓	↓	↓	↓	↓	↓	↓	↓	↓	↓	↓	↓	0 1	↓	↓	1 2	2 3	3 4	5 6	8 9	12 13	18 19	27 28	41 42
C	5	↓	↓	↓	↓	↓	↓	↓	↓	↓	↓	↓	↓	↓	0 1	↓	↓	1 2	2 3	3 4	5 6	8 9	12 13	18 19	27 28	41 42	↑
D	8	↓	↓	↓	↓	↓	↓	↓	↓	↓	↓	↓	↓	0 1	↓	↓	1 2	2 3	3 4	5 6	8 9	12 13	18 19	27 28	41 42	↑	↑
E	13	↓	↓	↓	↓	↓	↓	↓	↓	↓	↓	↓	0 1	↓	↓	1 2	2 3	3 4	5 6	8 9	12 13	18 19	27 28	41 42	↑	↑	↑
F	20	↓	↓	↓	↓	↓	↓	↓	↓	↓	↓	0 1	↓	↓	1 2	2 3	3 4	5 6	8 9	12 13	18 19	↑	↑	↑	↑	↑	↑
G	32	↓	↓	↓	↓	↓	↓	↓	↓	↓	0 1	↓	↓	1 2	2 3	3 4	5 6	8 9	12 13	18 19	↑	↑	↑	↑	↑	↑	↑
H	50	↓	↓	↓	↓	↓	↓	↓	↓	0 1	↓	↓	1 2	2 3	3 4	5 6	8 9	12 13	18 19	↑	↑	↑	↑	↑	↑	↑	↑
J	80	↓	↓	↓	↓	↓	↓	↓	0 1	↓	↓	1 2	2 3	3 4	5 6	8 9	12 13	18 19	↑	↑	↑	↑	↑	↑	↑	↑	↑
K	125	↓	↓	↓	↓	↓	↓	0 1	↓	↓	1 2	2 3	3 4	5 6	8 9	12 13	18 19	↑	↑	↑	↑	↑	↑	↑	↑	↑	↑
L	200	↓	↓	↓	↓	↓	0 1	↓	↓	1 2	2 3	3 4	5 6	8 9	12 13	18 19	↑	↑	↑	↑	↑	↑	↑	↑	↑	↑	↑
M	315	↓	↓	↓	↓	0 1	↓	↓	1 2	2 3	3 4	5 6	8 9	12 13	18 19	↑	↑	↑	↑	↑	↑	↑	↑	↑	↑	↑	↑
N	500	↓	↓	↓	0 1	↓	↓	1 2	2 3	3 4	5 6	8 9	12 13	18 19	↑	↑	↑	↑	↑	↑	↑	↑	↑	↑	↑	↑	↑
P	800	↓	↓	0 1	↓	↓	1 2	2 3	3 4	5 6	8 9	12 13	18 19	↑	↑	↑	↑	↑	↑	↑	↑	↑	↑	↑	↑	↑	↑
Q	1250	↓	0 1	↓	↓	1 2	2 3	3 4	5 6	8 9	12 13	18 19	↑	↑	↑	↑	↑	↑	↑	↑	↑	↑	↑	↑	↑	↑	↑
R	2000	0 1	↑	↓	1 2	2 3	3 4	5 6	8 9	12 13	18 19	↑	↑	↑	↑	↑	↑	↑	↑	↑	↑	↑	↑	↑	↑	↑	↑
S	3150			1 2																							

↓ = Use first sampling plan below arrow. If sample size equals or exceeds lot or batch size, do 100 percent inspection.
↑ = Use first sampling plan above arrow.
Ac = Acceptance number.
Re = Rejection number.

FIGURE 7.8 Single Sampling Plans for Tightened Inspection (Master Table). = Use first sampling plan below. If sample size equals, or exceeds, lot or batch size, do 100 percent inspection. = Use first sampling plan above arrow. Ac = Acceptance number. Re = Rejection number.

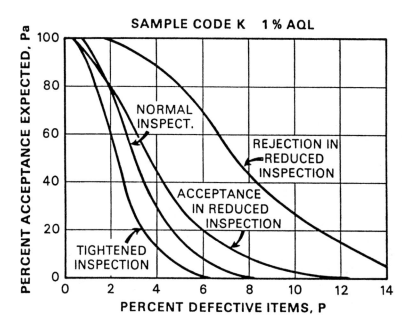

FIGURE 7.9   Inspection of Normal, Tightened, and Reduced OC Curves—Sample Code K 1% AQL.

for goods produced by continuous processing. Within MIL-STD-105D and MIL-STD-414, in addition to the single sampling plans discussed, there are also sampling plans for double and multiple sampling.

These latter types of sampling plans involve sequential sampling, using smaller first samples. Such sampling plans are economical to use when experience from previous inspections shows that the process is capable of producing lots that consistently have a number of defective or nonconforming parts lower than the required AQL. A table is provided so that the user can compare the average sample size for double and multiple sampling versus single sampling for a given process quality level.

Other sampling plans that may come to the reader's attention are:

1. Skip lot sampling;
2. Chain sampling;
3. Continuous sampling; and
4. Lot-plot sampling and numerous others.

Sample size code letter	Sample size	Acceptable Quality Levels (normal inspection)													
		.04	.065	.10	.15	.25	.40	.65	1.00	1.50	2.50	4.00	6.50	10.00	15.00
		k	k	k	k	k	k	k	k	k	k	k	k	k	k
B	3	↓	↓	↓	↓	↓	↓	↓	▼	▼	1.12	.958	.765	.566	.341
C	4	↓	↓	↓	↓	↓	↓	↓	1.45	1.34	1.17	1.01	.814	.617	.393
D	5	↓	↓	↓	↓	↓	↓	1.65	1.53	1.40	1.24	1.07	.874	.675	.455
E	7	↓	↓	↓	↓	2.00	1.88	1.75	1.62	1.50	1.33	1.15	.955	.755	.536
F	10	↓	↓	2.24	2.11	1.98	1.84	1.72	1.58	1.41	1.23	1.03	.828	.611	
G	15	2.64	2.53	2.42	2.32	2.20	2.06	1.91	1.79	1.65	1.47	1.30	1.09	.886	.664
H	20	2.69	2.58	2.47	2.36	2.24	2.11	1.96	1.82	1.69	1.51	1.33	1.12	.917	.695
I	25	2.72	2.61	2.50	2.40	2.26	2.14	1.98	1.85	1.72	1.53	1.35	1.14	.936	.712
J	30	2.73	2.61	2.51	2.41	2.28	2.15	2.00	1.86	1.73	1.55	1.36	1.15	.946	.723
K	35	2.77	2.65	2.54	2.45	2.31	2.18	2.03	1.89	1.76	1.57	1.39	1.18	.969	.745
L	40	2.77	2.66	2.55	2.44	2.31	2.18	2.03	1.89	1.76	1.58	1.39	1.18	.971	.746
M	50	2.83	2.71	2.60	2.50	2.35	2.22	2.08	1.93	1.80	1.61	1.42	1.21	1.00	.774
N	75	2.90	2.77	2.66	2.55	2.41	2.27	2.12	1.98	1.84	1.65	1.46	1.24	1.03	.804
O	100	2.92	2.80	2.69	2.58	2.43	2.29	2.14	2.00	1.86	1.67	1.48	1.26	1.05	.819
P	150	2.96	2.84	2.73	2.61	2.47	2.33	2.18	2.03	1.89	1.70	1.51	1.29	1.07	.841
Q	200	2.97	2.85	2.73	2.62	2.47	2.33	2.18	2.04	1.89	1.70	1.51	1.29	1.07	.845
		.065	.10	.15	.25	.40	.65	1.00	1.50	2.50	4.00	6.50	10.00	15.00	
		Acceptable Quality Levels (tightened inspection)													

All AQL values are in percent defective.
↓ Use first sampling plan below arrow, that is, both sample size as well as k value. When sample size equals or exceeds lot size, every item in the lot must be inspected.

FIGURE 7.10.   Standard Deviation Method Master Table for Normal and Tightened Inspection for Plans Based on Variability Unknown (Single Specifications Limit-Form 1). All AQL values are in percent defective. Use first sampling plan below arrow, that is, both sample size as well as K value. When sample size as well as K value. When sample size equals or exceeds lot size, every item in the lot must be inspected.

One additional sampling scheme of importance, since it is the first of the many used by industry, is the Dodge-Rohmig "Sampling Inspection Tables," which provide four sets of tables: single and double sampling plans, based on lot tolerance percent defective (LTPD), and single and double sampling plans based on the AOQL. Both sets of plans minimize the amount of inspection required. These sampling plans were developed at the Bell Telephone Laboratories by Drs. Dodge and Rohmig, primarily to check the quality of goods and equipment shipped between Western Electric plants. These are considered to be "Consumer's Risk" sampling plans, and there are two important considerations involved in their use. First, it is assumed that all rejected lots will be subjected to 100% screening. Secondly, in order to use the tables, which must be entered by the "Process Average Percent," it is necessary to have some prior knowledge about the quality of the product being inspected. If this knowledge is not available, it must be obtained from the supplier, or estimated. Such estimations can lead to dangerous error. This limitation on the use of Dodge-Rohmig plans is in contrast to the military and ANSI/ASQC standards discussed earlier, which may be used without any prior knowledge of the process average.

References

Dodge, H. F., and Harry G. Romig. 1944. Sampling Inspection Tables. New York: John Wiley & Sons, Inc.

Duncan, Acheson J. 1955. Quality Control and Industrial Statistics. Homewood, IL: Richard D. Irwin, Inc.

Grant, E. L., and R. S. Leavenworth. 1972. Statistical Quality Control, 4th Ed. New York: McGraw-Hill Book Co.

1963. Sampling Procedures and Tables for Inspection by Attributes—MIL-STD-105D. Washington, D. C.: Department of Defense.

1957. Sampling Procedures and Tables for Inspection by Variables for Percent Defective—MIL-STD-414. Washington, D. C.: Department of Defense.

1965. Guide for Sampling Inspection—MIL-HDBK-53. Washington, D. C.: Department of Defense.

1980. American National Standard—Sampling Inspection and Tables for Inspection by Variables for Percent Non-conforming. ANSI/ASQC Z1.9 1980. Milwaukee, WI: American Society for Quality Control.

1981. American National Standard—Sampling Inspection and Tables for Inspection by Attributes—ANSI/ASQC Z1.4-1981. Milwaukee, WI: American Society for Quality Control.

8

Correlation and Regression

Correlation and Regression Analysis can be used for a number of purposes that are of interest to the laboratory practitioner. It is often useful to know about the relationship between two variables and how they are dependent on each other. This dependency, when established, permits us to forecast how one process will perform, by observing the results of another. This technique is useful when one test is destructive or the results are difficult to obtain directly. Many analytical and testing methods that are easily applied or are nondestructive are highly correlated with like destructive tests. This being true, we can use substitute methods, with a reasonable degree of confidence that we will arrive at the sought after results.

Linear Regression is a mathematical solution to the plot of a straight line through a scatter diagram (Figure 8.1). Simple Correlation "r" is a measurement of how well that straight line represents the scatter diagram. In Figure 8.1, we have five different scatter diagrams, displaying various degrees of scatter. A distribution of data points as shown in diagram (a) is said to be heteroscedastic or widely scattered, while distributions pictured in (c) and (d) are said to be homoscedastic, since they fit the straight line pattern more closely.

The "best fit" straight line drawn in a scatter diagram may have either a negative or positive slope.

A positively sloped line represents positive correlation, and a negative slope represents, of course, negative correlation. Correlation alone does not necessarily determine the cause and effect relationship, which should be determined some other way. The fact that the relationship has correlation, positive or negative, only says that the two variables move in relationship to each other. To plot the line for a given set of data, we compute to the Least Squares Regression Line in the following manner.

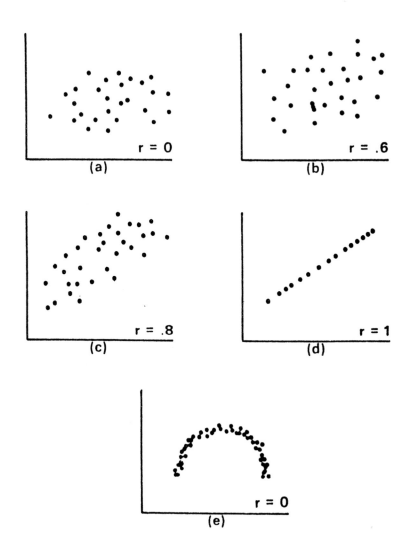

FIGURE 8.1.   Scatter Diagrams.

The Least Squares Line is a line for which the sum of the distances, from the line to the individual data points squared, is a minimum.

Figure 8.2 is a representation of a Least Squares Line plot.

In simple regression, linear regression is always represented by a straight line. In mathematics, any straight line can be represented by an equation, $Y = a + bx$, where:

DIAMETER VS. HEIGHT OF TREES

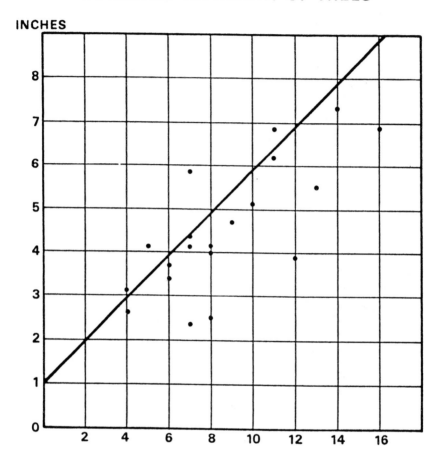

FIGURE 8.2.   Diameter vs. Height of Trees.

Y_c = a calculated point

  a = a constantt

  b = the slope of the line

  x = a given value

Let the data be represented by $(X_1, Y_1,)$ (X_2, Y_2).......

(X_n, Y_n).

$\hat{Y} = a + bX$

$\hat{Y}$ is the predicted Y value for the corresponding given X value.

The first step is to change the equations to normal form as follows:

1. $\Sigma Y = Na + b \Sigma X$; then, multiplying through by ΣX
2. $\Sigma XY = a \Sigma X + b \Sigma X^2$

to arrive at two simultaneous equations, where N = the number of data pairs. To illustrate how these equations can be used, we will examine the relationship of tree trunk diameters to their heights, knowing that if we can establish a correlation between the height of the trees and the diameter of the trunk, we can estimate the height of a tree by finding the diameter of its trunk. We use this example because the concept of the relationship between the trunk diameter and the tree height is easy to understand.

The next step in fitting a line to actual data is to list all the measured heights and corresponding diameters and their products (XY) and squares (X^2 and Y^2), and then total the data columns (Table 8.1), to arrive at the summations of X, Y, XY, X^2, and Y^2. This data is then entered in the equations above:

TABLE 8.1

X (Diameter)"	Y (Height)·	XY	X^2	Y^2
1. 2.3	7	16.1	5.29	49
2. 2.5	8	20.0	6.25	64
3. 2.6	4	10.4	6.76	16
4. 3.1	4	12.4	9.61	16
5. 3.4	6	20.	11.56	36
6. 3.7	6	22.2	13.69	36
7. 3.9	12	46.8	15.21	144
8. 4.0	8	32.0	16.00	64
9. 4.1	5	20.5	16.81	25
10. 4.1	7	28.7	16.81	49
11. 4.2	8	33.6	17.64	64
12. 4.4	7	30.8	19.36	49
13. 4.7	9	42.3	22.09	81
14. 5.1	10	51.0	26.01	100
15. 5.5	13	71.5	30.25	169
16. 5.8	7	40.6	33.64	49
17. 6.2	11	68.2	38.44	121
18. 6.9	11	75.9	47.61	121
19. 6.9	16	110.4	47.61	256
20. 7.3	14	102.2	53.29	196
Σ 90.7	173	856.0	453.93	1,705

1. $173 = 20a + 90.7b$
2. $856.0 = 90.7a + 453.93b$

Then, solving for b in Equation 2.

2. $nl = 856.0 = 90.7a + 453.93b$
3. $784,555 = 90.7a + 411.3245b$ (Equation 1. times 4.535)

(4.535 is the mean diameter of the trees, 90.7/20 = 4.535) Now subtract 3. from 2., to get:

$71.445 = 42.6055b$
$b = 71.445/42.6055 = 1.676896$

Now in equation 1. solve for a;

$Y = Na + Xb$
$173 = 20a + (90.7)(1.676896)$
$173 = 20a + 152.094467$
$20a = 173 - 152.094467$
$a = 1.045277$

When performing these calculations, it is wise not to round off too early, carrying out computations to at least three decimal places.

We now have values for a and b, which we can use in our original equation describing the regression line:

$Y = a + bX$

substituting our calculated values for a and b as follows:

$Y = 1.045 + 1.677X$

This will produce the Least Squares Regression Line (Figure 8.2). If the diameter of a tree is substituted for the value X, then Y will be the estimated height of the tree. This is an example of applying Linear Regression.

In this example, we have been dealing with independent variables, the height of the tree and the diameter of the trunk, and we use the information from one that we can measure, to get an estimate of the other. The question then arises: How close will the estimate be? When dealing with least squares, we use the Standard Error of the Estimate (SY.X), instead of the standard deviation, as a measurement of the variability of the data points around the regression line. The standard error of the estimate can be calculated in much the same way that a standard deviation is calculated. It represents information about the goodness of fit of the regression line, in the same manner as the standard deviation tells us about the the spread of data around the mean.

To arrive at the standard error of the estimate, we add to our table two more columns: Y - Y and $(Y - Y)^2$. Y is the tree height and Y is obtained by substituting the tree diameters (X) from the first column in the table (Table 8.2) in the equation Y = a + bx.

The next step is to use the equation.

$$SY. X = \sqrt{\frac{\Sigma(Y - Y_c)^2}{N}}$$

to calculate the standard error of the estimate. Note that the equation is very

TABLE 8.2

	X (Diameter)"	Y (Height)'	XY	X^2	Y^2	Y-Yc	$(Y-Yc)^2$
1.	2.3	7	16.1	5.29	49	2.098	4.4016
2.	2.5	8	20.0	6.25	64	2.762	7.6286
3.	2.6	4	10.4	6.76	16	-1.405	1.9740
4.	3.1	4	12.4	9.61	16	-2.244	5.0355
5.	3.4	6	20.4	11.56	36	-0.747	0.5580
6.	3.7	6	22.2	13.69	36	-1.250	1.5625
7.	3.9	12	46.8	15.21	144	4.415	19.4922
8.	4.0	8	32.0	16.00	64	0.247	0.0610
9.	4.1	5	20.5	16.81	100	-2.921	8.5322
10.	4.1	7	28.7	16.81	49	-0.921	0.8482
11.	4.2	8	33.6	17.64	64	-0.088	0.0077
12.	4.4	7	30.8	19.36	49	-1.424	2.0278
13.	4.7	9	42.3	22.09	81	0.073	0.0053
14.	5.1	10	51.0	26.01	100	0.402	0.1616
15.	5.5	13	71.5	30.25	169	2.732	7.4638
16.	5.8	7	40.6	33.64	49	-3.772	14.2280
17.	6.2	11	68.2	38.44	121	-0.442	0.1954
18.	6.9	11	75.9	47.61	121	-1.616	2.6115
19.	6.9	16	110.4	47.61	256	3.384	11.4515
20.	7.3	14	102.2	53.29	196	0.713	0.5084
TOTALS	90.7	173	856.0	433.93	1,705	-0.004	88.7548

similar to that for the standard deviation. Substituting the calculated data in the equation above, we have:

$$SY.X = \sqrt{88.75/20} \ = \ \sqrt{4.438} \ = 2.107$$

Now that we have determined the standard error of the estimate, the next step is to use this information, to obtain the Correlation Coefficient. The correlation coefficient is used to judge to what degree our two independent variables are correlated. The correlation coefficient is a relative measure mathmatically calculated, but subjectively evaluated. If the trees get taller and the diameters increase correspondingly, this is positive or direct correlation. If the relationship is in the other direction, the correlation is called negative or inverse correlation. Since, in this case, the slope of the regression line is positive, we can expect that the result of our calculation for r, the correlation coefficient, will be positive. The equation for r is:

$$r = \frac{N \, \Sigma XY - (\Sigma X)(\Sigma Y)}{\sqrt{[N \, \Sigma X^2 - (\Sigma X)^2][N \, \Sigma \, Y^2 - (\Sigma Y)^2]}}$$

Substituting from our solution table (Table 8.2):

$$r = \frac{20)(856.0) - (90.7)(173)}{\sqrt{[20)(453.93) - (90.7)^2][(20)(1705) - (173)^2]}}$$

$$r = \frac{(17120) - (15691.1)}{\sqrt{[(9078) - (8226.49)][(34100) - (29929)]}}$$

$$r = \frac{1428.9}{\sqrt{(851.51)(4171)}} \qquad = \frac{1428.9}{\sqrt{3551584.21}}$$

$$r = \frac{1428.9}{1884.582} \qquad = +0.758$$

Note the positive sign for r.

HOW TO EVALUATE R

The correlation coefficient is a measure of the degree with which the independent variable and its partner move either together or in opposition. A positive result indicates direct correlation. Conversely, a negative result would indicate inverse correlation.

When r = O, there is no correlation, when r = 1.0, there is perfect correlation, thus, the higher the coefficient and the greater correlation.

There is no absolute, subjective reading that can be attached to any given coefficient result. However, since laboratory results are usually the output of precision measurement methods, a rule of thumb can be used for evaluating the correlation coefficient as follows:

.00–.25	Doubtful correlation
.26–.50	Fair correlation
.51–.75	Good correlation
.76–1.00	Superior correlation

These guidelines are very general in nature and should be used only to gain an understanding of how r values may be interpreted. For different fields of application, more specific guidelines should be developed. For instance, the correlation of environmental exposure to a specific pollutant, with some adverse health condition, may never result in a high correlation coefficient because of all the uncontrolled variables affecting the results.

A caveat is in order at this time. It is not necessarily true that a relationship measured by r is meaningful. The sample on which the data is based must be large enough to ensure that the influence of chance causes of variation is minimized. Secondly, there must be a rational relationship of the two variables under investigation. Following is an example of what must be avoided: Over a period of years, the correlation coefficient between the rise in teachers' salaries and the consumption of hard liquor in the United States was found to be 0.98.

References

Duncan, A. J. 1955. Quality Control and Industrial Statistics. Homewood, IL: Richard D. Irwin, Inc.

Schrock, E. M. 1957. Quality Control and Statistical Methods. New York: Reinhold Publishing Corporation.

Snedecor, G. W., and W. G. Cochran. 1956. Statistical Methods. Ames, IA: The Iowa State College Press.

9

Introduction to Outlying Observations

INTRODUCTION

Dealing with outlying observations or "outliers" is one of the most frustrating and perplexing problems facing laboratory workers. The writings of some "strict" statisticians would lead us to a position of "never rejecting an observation," unless a physical cause can be documented. In that case, statistics aren't needed anyway. The other end of the spectrum is the temptation to reject all "bad looking" or deviant observations, summarily. We must operate between these extremes, to assure a clear conscience and a statistically and scientifically defensible program. After looking at some historical background, this chapter will discuss the philosophical problems, the scope of possible potential problems, and the use of hypothesis testing.

As noted, outliers may be rejected for some known or discovered physical reason, such as the detection of an out-of-calibration instrument or for some yet to be identified mechanical or method aberation. The other reason for rejection is the determination by statistical test that an outlier is suspected to be present.

In dealing with outliers, the shape of the distribution of the raw data must be considered. When applying tests for outlying data points, it may be necessary to transform the data (using the logarithms of such data points to arrive at a normal distribution, for instance).

BACKGROUND

One cannot, without attendant risk, assume that a set of observations are indeed a random sample from a population. Shewhart (1940), at a University of Pennsylvania Conference, gave practical criteria. One should consider a measurement process to be (i.e., should "decide to act for the present as if" the process were) in a state of (simple) statistical control, only if the measurements at hand show no evidence of a lack of statistical control when the data are analyzed for randomness, in the order in which they were taken by the control chart techniques, for averages and standard deviations that Shewhart had found so valuable in industrial process control. Or certain additional tests for randomness based on "runs above and below average" and "runs up and runs down" can be used.

Attaining an ideal strict statistical control may be difficult, if not impossible, to achieve. The economic feasibility may negate practicability, even if it is technically possible. However, the observer or data analyst assumes greater and greater risks in accepting the assumptions that underlie the tests for outliers, the further the departure from statistical control.

OVERVIEW OF OUTLIER PROBLEMS

The process of detecting outliers is used for one of three purposes:

1. To screen routine data sets prior to statistical analysis (data validation);
2. To determine if the process being monitored has changed significantly in mean or variance; or
3. To detect extreme values that may result from improvements due to process or method changes.

Most of the traditional outlier tests have been devised for looking at a single population. Slippage tests have been developed to test the equality of several populations with the alternative hypothesis that one of the populations is different or has "slipped." When each population is defined by only a single observation, the classical detection of outlier techniques apply to this limiting case.

Purposes 2. and 3. can be related to process quality control and investigations. Fleming's discovery of penicillin is such a case. Screening a set or sets of data prior to statistical analysis (data validation) is the usual purpose for which outlier tests are put into use. Table 9.1 illustrates three common situations.

Case A and Case B seem to allow clear-cut decisions intuitively. Case C, of course, bothers us the most. Before we worry about "how" to deal with this case, we should consider the "why." What is the cause of the suspected outlier? What is the nature of the population from which this situation has developed?

Figure 9.1 shows two potential populations. The situations described in Figure

Table 9.1   Examples of Data Sets

Case A	00000		0
Case B	000000		
Case C	00000	0	

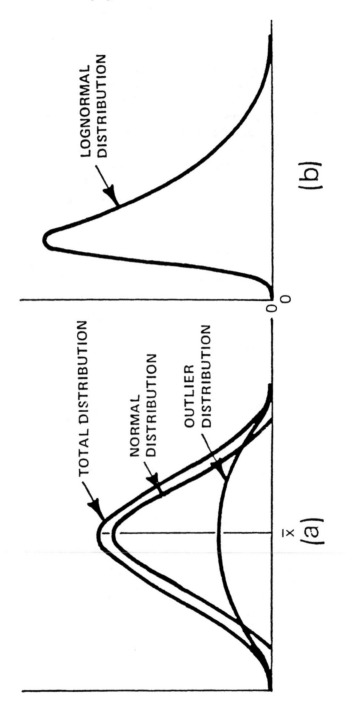

FIGURE 9.1.   Population Distributions.

9.1 are both routinely encountered in the laboratory. Case (a) may arise from equipment or other malfunctions of the sampling or analytical procedures of the measurement process. Case (b) represents the distribution of emission or exposure values, in the workplace environment. Much of the work of the National Institute for Occupational Safety and Health (NIOSH), published in its Occupational Exposure Sampling Strategy Manual, is based on this assumption.

It is well to recall that the preferred technique is to determine the physical or chemical cause of the outlier observation and, from this, make the rejection decision.

Typical current publications of the American Chemical Society (ACS) and the American Society for Testing and Materials (ASTM-1988) may acknowledge the need to handle outliers, but provide little specific guidance, so the classical methods are still being used. Therefore, these basic references will be utilized in this and the following chapters on outliers.

Regardless of the protocol used for rejecting outliers, it is important to acknowledge their presence in the original data and describe the rejection criteria. "It is much better to state their values and to do alternative analysis using all or some of them." (Kruskal 1960) Another caution relates to the ultimate use of the data being generated. If, after performing an analytical or testing procedure, one of, say, five measurements is obviously out of line, we would reject it. Such rejection, especially if a cause such as equipment malfunction is found, can be justified. However, if we are interested in determining the effectiveness of the monitoring procedure, the rejected observation and its cause are a vital part of the process information package.

The warning issued by Pearson and Chandrasehar in 1936 and quoted by K. R. Nair in 1948 bears repeating here.

> To base the choice of the test of a statistical hypothesis upon an inspection of the observations is a dangerous practice; a study of the configuration of a sample is almost certain to reveal some feature, or features which are exceptional if the hypothesis is true. By choosing the feature most unfavorable to the hypothesis out of a very large number of features examined, it will usually be possible to find some reason for rejecting the hypothesis.

BASE POPULATION CHARACTERISTICS

In the ideal world, the variability associated with our measurement processes would be described by the Normal or Gaussian distribution discussed in earlier chapters. In analytical or testing work, many measurement situations are better described (at least empirically) by the log-normal distribution. (Figure 9.1) It is not necessary to know the causes of this situation, even though that would assist in understanding a data set and in dealing with outliers. In a log-normal distribution, one can assume that the lower end of the curve is truncated by zero or at least by the lower limit of detection of the measurement method.

Other population distributions might be observed as well. The practical problem in outlier rejection is that the validity or efficiency of the test is different for differently shaped populations. There are at least three approaches that can be used in dealing with this problem. The simplest involves the Central Limit Theorem, which has been described by the following:

If a population has a finite variance σ^2 and the mean μ, then the distribution of the sample mean (of n independent measurements) approaches the normal distribution with variance σ^2/n and mean m as sample size n increases.
This remarkable and powerful theorem is indeed tailored for measurement processes. (Ku)

The central limit theorem thereby provides an effective normalization procedure for data, even when the population distribution is markedly non-normal.

The second approach is to use normalizing transformations. When we know, from previous experience or on a theoretical basis, that a transformation will approximately normalize the data set, it may be more efficient and convenient to use normal-based outlier tests than distribution free tests. (Natrella, 1963)

The third approach is to use distribution free tests for outliers. One such approach is to use ranks (Wonnacott 1977 or Conover 1971).

THE CHOICE OF THE OUTLIER REJECTION TEST

The choice of which test to apply, to examine a set of data for rejection of a single outlier, is based on several considerations. The first consideration is whether the base population is known or assumed to be normally distributed. The following listings of tests (Table 9.2) are based on that assumption. The second consideration is how much is known about μ and σ. The third consideration is whether we are interested in shifts in a specified direction (one-sided) or in either direction (two-sided). Since in the majority of testing or analytical laboratory applications, the suspect outlier undergoing scrutiny will be on the right or upper side, this formulation is given for one-sided tests.

The list of tests in Table 9.2 is based on David, pp. 172–173 (1970). The choice between available tests that are appropriate to the sets of conditions shown above will be discussed in the following chapters.

OUTLIER TESTS WHEN μ AND σ ARE KNOWN

The case where both μ and σ are known is a basic theoretical situation. Laboratory practitioners will encounter this set of circumstances rarely, if at all, in routine operations; therefore, it will not be discussed here.

The first and simplest test is based on using the Cumulative Normal Distribution

Table 9.2.   Tests for Outliers

One-Sided	Two-Sided

When both μ are σ are known

$$A_1 = \frac{(X_{(n)} - \mu)}{\sigma}$$

$$A_2 = \max \frac{|X_i - \mu|}{\sigma}$$

Pearson & Hartley 1966
Table 24

$$A_3 = x^{2_n} = \frac{\Sigma(X_i - \mu)^2}{\sigma^2}$$

When only σ is known

$$B_1 = \frac{(X_{(n)} - \overline{X})}{\sigma}$$

Grubbs (1950)

$$B_2 = \max \frac{|X_1 - \overline{X}|}{\sigma}$$

Pearson & Hartley 1966
Table 22

$$B_3 = \frac{(X_{(n)} - X_{(n-1)})}{\sigma}$$

$$B_4 = \frac{Range}{\sigma}$$

$$B_5 = x_{n-1}^2 = \frac{\Sigma(X_1 - \overline{X})^2}{\sigma^2}$$

Both μ and σ unknown but have an Independent Estimate of σ^2

$$C_1 = \frac{(X_n - \overline{X})}{S_v}$$

$$C_2 = \max \frac{|X_i - \overline{X}|}{S_v}$$

Where S_v^2 is an independent mean square estimate of σ^2 based on V Degrees of Freedom.

Pearson & Hartley
Table 26

Halperin et. al. 1955

$$C_3 = \frac{Range}{S_v}$$

Pearson Hartley
Table 29

$$C_4 = \frac{(X_{(n)} - \overline{X})}{[\Sigma(X_1 - X)^2 + VS_v^2]^{1/2}}$$

$$C_5 = \frac{\max |X_1 - \overline{X}|}{[\Sigma(X_i - X)^2 + VS_v^2]^{1/2}}$$

Pearson & Hartley
Table 26a

Pearson & Hartley
Table 26b

Table 9.2

One-Sided	Two-Sided

Both μ and σ Unknown

$$D_1 = \frac{(X_{(n)} - \overline{X})}{\dfrac{[\Sigma(X_i - \overline{X})^2]^{1/2}}{n}}$$

Grubbs 1950

$$D_2 = \frac{\text{Max } |X_i - \overline{X}|}{\dfrac{[\Sigma(x_i - \overline{x})^2]^{1/2}}{n - 1}}$$

Pearson & Hartley
Table 26b

$$D_3 = \frac{\text{Range}}{\dfrac{[\Sigma(X_i - \overline{X})^2]^{1/2}}{n - 1}}$$

Pearson & Hartley
Table 26b

$$D_4 = \frac{n^{1/2}\Sigma(X_i - \overline{X})^3}{[\Sigma(X_i - \overline{X})^2]^{3/4}}$$

$$D_5 = \frac{n\Sigma(X_i - \overline{X})^4}{[\Sigma(X_i - X)^2]^2}$$

Pearson & Hartley
Table 34c

$$D_6 = \frac{\displaystyle\sum_{i=1}^{n-2}\left[X_{(i)} - \left(\frac{\displaystyle\sum_{i=1}^{n-2} X_{(i)}}{n - 2}\right)\right]^2}{\displaystyle\sum_{i=1}^{n}(X_i - \overline{X})^2}$$

Grubbs, 1950

D_7 = Dixon's r statistics, Dixon 1951

Values of z_p corresponding to P for the normal curve.

z is the standard normal variable

P	.00	.01	.02	.03	.04	.05	.06	.07	.08	.09
.00	—	−2.33	−2.05	−1.88	−1.75	−1.64	−1.55	−1.48	−1.41	−1.34
.10	−1.28	−1.23	−1.18	−1.13	−1.08	−1.04	−0.99	−0.95	−0.92	−0.88
.20	−0.84	−0.81	−0.77	−0.74	−0.71	−0.67	−0.64	−0.61	−0.58	−0.55
.30	−0.52	−0.50	−0.47	−0.44	−0.41	−0.39	−0.36	−0.33	−0.31	−0.28
.40	−0.25	−0.23	−0.20	−0.18	−0.15	−0.13	−0.10	−0.08	−0.05	−0.03
.50	0.00	0.03	0.05	0.08	0.10	0.13	0.15	0.18	0.20	0.23
.60	0.25	0.28	0.31	0.33	0.36	0.39	0.41	0.44	0.47	0.50
.70	0.52	0.55	0.58	0.61	0.64	0.67	0.71	0.74	0.77	0.81
.80	0.84	0.88	0.92	0.95	0.99	1.04	1.08	1.13	1.18	1.23
.90	1.28	1.34	1.41	1.48	1.55	1.64	1.75	1.88	2.05	2.33

Special Values

P	.001	.005	.010	.025	.050	.100
z_p	−3.090	−2.576	−2.326	−1.960	−1.645	−1.282

P	.999	.995	.990	.975	.950	.900
z_p	3.090	2.576	2.326	1.960	1.645	1.282

FIGURE 9.2   Cumulative Normal Distribution—Values of zp. Values of z_p corresponding to P for the normal curve z is the standard normal variable. Special values

values of Figure 9.2, taken from NIST Handbook 91, page T-3, (Natrella 1963) and other textbooks. The test is described in Natrella, Chapter 17.

The second test involves using the Chi-square χ^2 test (Natrella, Ch. 9). The specific comparison would be that of comparison to a theoretical standard (the cumulative normal distribution). Several categories are used to describe the population characteristics. This test is, of course, usually used for a broader test for normality, as we learned earlier in Chapter 3, and its use to detect outliers is just one application. Then one uses a minimum number of cells, such as three, for a one-sided test.

Using a two-sided test would require the cells to include one value at each tail of the distribution.

SUMMARY

This chapter has reviewed the philosophy of outlier rejection, as well as the practical limitations. Outlier rejection techniques have been derived from both historical and theoretical standpoints. A survey of selected techniques illustrated the limitations, based upon the assumption of a normal population and differing levels of information about the population estimates of μ and σ. Several techniques for outliers illustrative of the most knowledge (both μ and σ known) apply the principles covered earlier.

References

Conover, W. J. 1971. *Practical Nonparametric Statistics*. New York: John Wiley & Sons, Inc.

David, H. A. 1970. *Order Statistics*. New York: John Wiley & Sons, Inc.

Grubbs, F. E. 1950. Sample criteria for testing outlying observations. *Annals of Mathematical Statistics*. 21:27-58.

Halperin, M., S. W. Greenhouse, J. Cornfield, and J. Zalahor. 1950. Tables of percentage points for studentized maximum deviate in normal samples. *Journal of the American Statistical Association*. 50:185-95.

Irwin, J. O. 1925. On a criteria for the rejection of outlying observations. *Biometrika* 17:238-50.

Kruskal, William H. Some remarks on wild observations. *Technometrics*. 2:1960.

Ku, Harry H. 1969. Statistical concepts in metrology. In *Precision Measurement and Calibration—Statistical Concepts and Procedures*, ed. Harry H. Ku, Vol. 1, NIST Special Publication 300, Washington, DC: NIST.

Leidel, N. A., K. A. Busch, and J. R. Lynch. 1977. *Occupational Exposure Sampling Strategy Manual*. Cincinnati: National Institute for Occupational Safety and Health.

Natrella, Mary. 1963. *Experimental Statistics*. Washington, DC: National Bureau of Standards.

Nair, K. R. 1948. On a criteria for the rejection of the extreme deviate from the sample mean and its studentized forms. *Biometrika* 35:114-118.

Pearson, E. S. and H. O. Hartley. 1966. *Biometrika Tables for Statisticians*, Vol. 1, 3rd Ed. Cambridge, England: Cambridge University Press.

Shewhart, Walter A. 1939. *Statistical Method from the Viewpoint of Quality Control.* Washington, DC: The Graduate School, U. S. Department of Agriculture.

Triegel, Elly K. 1988. Sampling variability in soils and solid wastes, principles of environmental sampling, in *ACS Professional Reference Book*, ed. H. Keith, Chapter 27. Washington, DC: American Chemical Society.

Wonnacott, T. H. and R. J. Wonnacott. 1977. *Introductory Statistics*. New York: John Wiley and Sons, Inc.

——. 1989. Atmospheric analysis; occupational health and safety. In *1989 Annual Book of ASTM Standards*, Vol. 11.03 Philadelphia: American Society for Testing and Materials.

——. 1980. *ASTM Standard P.practice for Dealing with Outlying Observations. E178-80.* Philadelphia: American Society for Testing and Materials.

10

Single Outliers

INTRODUCTION

The initial discussion of outliers set the stage for the problems of dealing with a suspect observation in a set of data. This set of data may be a sampling from a process where the variability has been so well established over a period of time that the population variability σ can be said to be known. A second situation occurs when we have less knowledge about the characteristics of the population. Information about the variability is not so certain, but we do have available an independent estimate of S_v. The second case is more common in laboratory operations, where we have established reasonably well the variability of a direct reading instrument, in similar situations. The third case is common also, where we have a set of data from a new plant, a new process, or a new analytical or testing procedure, without any historical data to assist in the interpretation. Both μ and σ are unknown.

This chapter will deal with the cases "A," where μ is unknown and σ known; "B", where σ is unknown, but there is an independent estimate, S_v of σ; and "C", where both μ and σ are unknown. As the title suggests, only single outliers will be dealt with in this chapter. This single outlier may occur at either end of the distribution, involving a two-sided test, or it may occur only at the right of the data set (the most common situation) or at the left side of the data set. These are one-sided tests.

Prior to discussing the several tests available for each of these situations, it would be good to reflect upon the types of events that cause the outlier that we are testing for, or what conditions have contaminated the measurement process, to cause the suspected outlier.

POPULATION CONTAMINATION

We will start with the assumption that the data set is a sample from a single normal population. Dixon (1953) has provided a good characterization of the problem. The normal population has a mean μ and a variance σ^2 and is described as follows: $N(\mu, \sigma^2)$. Suppose that the population distribution is at some point disturbed due to a calibration shift, for example. The mean μ is shifted by an increment described in terms of standard deviations from the mean or λ σ where λ is the number of units of displacement. After such a shift has occurred, the population is then described by: $N(\mu + \lambda \sigma, \sigma^2)$. A measurement or observation from this new population is called a location error. One could also experience an increase in the variability of the measurement process. A measurement or error from the more variable population $N(\mu, \lambda^2 \sigma^2)$ introduced into the original population $N(\mu, \sigma^2)$ is called a scalor error by Dixon. We have added this dimension to the discussion since we will need to evaluate the efficiency of tests for outliers, with respect to how well they perform when dealing with populations contaminated by either a displacement or scalor error.

WHEN ONLY σ IS KNOWN

The test B_1 and its derivatives were attributed to Grubbs (1950) and should also be attributed to Nair (1948). The test B_1 takes the form:

$$B_1 = \frac{(X_n - X)}{\sigma}$$

(10-1)

The two-sided version of this test, B_2, takes the form:

$$B_2 = \max \frac{[X_1 - X]}{\sigma}$$

(10-22)

Nair points out that the Irwin (1925) test statistic B_3, which uses the ratio:

$$B_3 = \frac{(X_n - X_{n-1})}{\sigma}$$

(10-3)

is not intuitively as good as the original form of B_1, as proposed by McKay in 1935. Nair also states that the range test for outliers proposed by Student in 1927, $\frac{(X_n - X_i)}{\sigma}$, or the B_4 form $\frac{Range\ (R)}{\sigma} = B_4$, will be the better test statistic if both X_1 and X_n are outliers. This case will be considered in the next chapter. David (1970) agrees with the conclusion of Nair.

The test statistic B5 or the χ^2 (Chi-square) test can also be a useful tool in outlier rejection.

Grubbs, in 1969, advocated his 1950 version of the B1 test statistic. Dixon, in 1950, tabulated the classification of outlier rejection tests, as presented earlier by David (1970). Dixon tested the various test statistics for both location and scalor errors. He concluded that test B_1 gives the best performance, if a single location error is present. The tests B_3 and B_4 are slightly easier to calculate, and their performance is nearly as good as that for B_1. However, when more than one outlier may be expected in a single sample, the performances of B_1 and B_4 increase, while B_3 decreases. Dixon's testing program produced the conclusion that test B_1 yields maximum performance, but that for ease of computation, almost equal performance for a single outlier, about equal performance for two outliers, and protection for outliers either above or below the mean, test statistic B_4 is recommended.

The tables to be used for test statistic B_1 are available either from Nair (Table 10.1—1948) or from Grubbs (Table 10.2—1950). The table for test statistic B_4 is provided by Pearson and Hartley (Table 10.3—1956).

Three example problems using the test statistic B_4 follow.

Example 10.1

A Tedlar bag with a concentration of carbon disulphide was prepared in a laboratory. Sixteen measurements were made of the contents. From previous experience, it is known that the population standard deviation is 2 ppm. The measurements were:

10.1	9.9	12.7	10.8
7.2	13.6	6.6	8.1
11.1	5.6	13.4	8.2
8.9	6.6	9.7	14.6

The test value is calculated, to determine if any of the above results should be rejected as outliers.

$$\text{Use test } B_4 = \underline{\text{range}} \qquad (10.4)$$

$$B_4 = \frac{R}{\sigma} \qquad \sigma = 2$$

$$B_4 = \frac{14.6 - 5.6}{2} = \frac{9}{2} = 4.5$$

The critical value (CV) for n = 20, from Table 10.3, for various confidence levels is:

| 10.0% | 5.0% | 2.5% | 1.0% | 0.5% | 0.1% |
| 4.69 | 5.01 | 5.30 | 5.65 | 5.89 | 6.41 |

Table 10.1   Test statistics $B_1 = \dfrac{(X_n - \overline{X})}{\sigma}$

Critical Values of B_1 When the Population Standard Deviation s is Known

Size of sample n	Lower percentage points						Upper percentage points					
	0.1	0.5	1.0	2.5	5.0	10.0	10.0	5.0	2.5	1.0	0.5	0.1
3	0.03	0.06	0.09	0.14	0.20	0.29	1.50	1.74	1.95	2.22	2.40	2.78
4	0.09	0.16	0.20	0.27	0.35	0.45	1.70	1.94	2.16	2.43	2.62	3.01
5	0.16	0.25	0.30	0.38	0.47	0.58	1.83	2.08	2.30	2.57	2.76	3.17
6	0.23	0.33	0.38	0.48	0.56	0.68	1.94	2.18	2.41	2.68	2.87	3.28
7	0.30	0.40	0.46	0.56	0.65	0.76	2.02	2.27	2.49	2.76	2.95	3.36
8	0.36	0.47	0.53	0.62	0.72	0.84	2.09	2.33	2.56	2.83	3.02	3.43
9	0.41	0.53	0.59	0.69	0.78	0.90	2.45	2.39	2.61	2.88	3.07	3.48

This table is taken from the paper of Nair 1948, page 140. By permission. Biometrika, Vol. 35, 1948.

Table 10.2   Test Statistic $B_1 = \dfrac{(X_n - \overline{X})}{\sigma}$

Critical Values of $B_{1(1)} = \dfrac{(\overline{X} - X_1)}{\sigma}$ or $B_{1\,n} = \dfrac{(X_n - \overline{X})}{\sigma}$

When the Population Standard Deviation σ is Known

Numbers of Observations n	5% Significance Level	1% Significance Level	0.5% Significance Level
2	1.39	1.82	1.99
3	1.74	2.22	2.40
4	1.94	2.43	2.62
5	2.08	2.57	2.76
6	2.18	2.68	2.87
7	2.27	2.76	2.95
8	2.33	2.83	3.02
9	2.39	2.88	3.07
10	2.44	2.93	3.12
11	2.48	2.97	3.16
12	2.52	3.01	3.20
13	2.56	3.04	3.23
14	2.59	3.07	3.26
15	2.62	3.10	3.29
16	2.64	3.12	3.31
17	2.67	3.15	3.33
18	2.69	3.17	3.36
19	2.71	3.19	3.38
20	2.73	3.21	3.39
21	2.75	3.22	3.41
22	2.77	3.24	3.42
23	2.78	3.26	3.44
24	2.80	3.27	3.45
25	2.81	3.28	3.46

This table is taken from the paper of Grubbs 1969, page 19. By permission.

Since all table (10.3) critical values are greater than the calculated or test value, no data points are rejected.

Example 10.2

Air samples from a certain factory area were collected over a period of time. From previous experience, it is known that the standard deviation of carbon monoxide

Table 10.3 Test Statistic $B_4 = \dfrac{\text{Range}}{\sigma}$

Percentage points of the distribution of the range

Size of sample n	Factor $1/d_n$	Lower percentage points						Upper percentage points					
		0·1	0·5	1·0	2·5	5·0	10·0	10·0	5·0	2·5	1·0	0·5	0·1
2	0.8862	0.00	0.01	0.02	0.04	0.09	0.18	2.33	2.77	3.17	3.64	3.97	4.65
3	.5908	0.06	0.13	0.19	0.30	0.43	0.62	2.90	3.31	3.68	4.12	4.42	5.06
4	.4857	0.20	0.34	0.43	0.59	0.76	0.98	3.24	3.63	3.98	4.40	4.69	5.31
5	.4299	0.37	0.55	0.66	0.85	1.03	1.26	3.48	3.86	4.20	4.60	4.89	5.48
6	0.3946	0.54	0.75	0.87	1.06	1.25	1.49	3.66	4.03	4.36	4.76	5.03	5.62
7	.3698	0.69	0.92	1.05	1.25	1.44	1.68	3.81	4.17	4.49	4.88	5.15	5.73
8	.3512	0.83	1.08	1.20	1.41	1.60	1.83	3.93	4.29	4.61	4.99	5.26	5.82
9	.3367	0.96	1.21	1.34	1.55	1.74	1.97	4.04	4.39	4.70	5.08	5.34	5.90
10	0.3249	1.08	1.33	1.47	1.67	1.86	2.09	4.13	4.47	4.79	5.16	5.42	5.97
11	.3152	1.20	1.45	1.58	1.78	1.97	2.20	4.21	4.55	4.86	5.23	5.49	6.04
12	.3069	1.30	1.55	1.68	1.88	2.07	2.30	4.29	4.62	4.92	5.29	5.54	6.09
13	.2998	1.39	1.64	1.77	1.97	2.16	2.39	4.35	4.68	4.99	5.35	5.60	6.14
14	.2935	1.47	1.72	1.86	2.06	2.24	2.47	4.41	4.74	5.04	5.40	5.65	6.19
15	0.2880	1.55	1.80	1.93	2.14	2.32	2.54	4.47	4.80	5.09	5.45	5.70	6.23
16	.2831	1.63	1.88	2.01	2.21	2.39	2.61	4.52	4.85	5.14	5.49	5.74	6.27
17	.2787	1.69	1.94	2.07	2.27	2.45	2.67	4.57	4.89	5.18	5.54	5.78	6.31
18	.2747	1.76	2.01	2.14	2.34	2.51	2.73	4.61	4.93	5.22	5.57	5.82	6.35
19	.2711	1.82	2.07	2.20	2.39	2.57	2.79	4.65	4.97	5.26	5.61	5.85	6.38
20	0.2677	1.87	2.12	2.25	2.45	2.62	2.84	4.69	5.01	5.30	5.65	5.89	6.41

The unit is the population standard deviation. Estimate of σ=range (or mean range) in a sample of n observations $\times 1/d_n$.

This table is taken from the volume of Pearson and Hartley 1956, page 156. By permission. Biometrika Tables for Statisticians, Vol. 1, Cambridge Press, England.

concentration levels has been 10 ppm. The results from this latest sampling were as follows:

CARBON MONOXIDE IN PPM

44	21	50	46	50
46	49	40	66	43

Should the value "66" be rejected as not coming from a normally distributed population?

Again, use test B_4

$$B_4 = \frac{Range}{\sigma} \quad \sigma = 10$$

$$B_4 = \frac{66 - 21}{10} = \frac{45}{10} = 4.5$$

$$n = 10$$

From Table 10.3, for n = 10, selected critical values are:

CV for 10% = 4.13
CV for 5% = 4.47
CV for 2.5% = 4.79

Since our calculated value is greater than the upper percentage point of 4.47 at the 5% percent confidence level, we conclude that the high result of 66 should be rejected. This set of values represents the type that may fit in the log-normal type of distribution.

Example 10.3

A laboratory technician reports that detector tube readings for carbon monoxide taken from the same factory area over a single day shift are as follows:

CARBON MONOXIDE IN PPM

48	51	47	49	50
52	50	51	49	50

These results seem to be remarkably consistent when compared to previous findings. The question is to determine whether this set of test results represents the data to be expected from such a series of measurements. This example illustrates the case where the variability of the data is too small to truly represent the population variability.

$$n = 10 \quad \sigma = 10$$

$$B_4 = \frac{\text{Range}}{\sigma} = \frac{5}{10} = .5$$

$CV_{0.1\%} = 1.08$
$CV_{0.5\%} = 1.38$
$CV_{1.0\%} = 1.47$

All of which are greater than the calculated value of .5.
Therefore, the data are rejected as being "too good."

BOTH μ AND σ UNKNOWN BUT HAVE AN INDEPENDENT ESTIMATE OF σ^2

This test situation involves knowledge that is less certain about the population variability. The test statistics tend to parallel the "B" test family of statistics. Test C_1 replaces the σ of B_1, with an independent estimate of σ or s_v. Test C_2 parallels test B_2. Test statistic C_3 or $C_3 = \text{Range}/s_v$ follows B_4. Test statistics C_4 and C_5 represent studentized versions of B_1.

The power of this series of tests is dependent upon the degrees of freedom (d.f.) involved in the calculation of s_v, As d.f. increases, their efficiency approaches that of test statistic B_1. Tables for a series of ratio test statistics have been compiled by Pearson and Hartley in 1956 (Table 10.4).

Two examples of how to use the test statistic C_1 and Table 10.5, associated with its use, follow.

Example 10.4

Ten respirable dust measurements were taken. The values obtained were:

RESPIRABLE DUST IN MG/M^3

7.55	9.74	8.48	14.58	9.53
15.33	9.00	5.91	6.88	7.25

Table 10.4a   Test Statistic C3 = $\dfrac{X_n - X_i}{S_v}$

Percentage points of the studentized range, $q = (x_n - x_1)/S_v$
Lower 5% points

v\n	2	3	4	5	6	7	8	9	10	11	12	13	14	15	16	17	18	19	20
10	0.09	0.43	0.75	1.01	1.20	1.37	1.52	1.63	1.74	1.82	1.91	1.98	2.05	2.12	2.17	2.22	2.26	2.30	2.34
11	.09	.43	.75	.01	.21	.38	.52	.64	.75	.84	.92	2.00	.07	.13	.18	.24	.28	.33	.37
12	.09	.43	.75	.01	.21	.38	.53	.65	.76	.85	.93	.01	.08	.14	.20	.26	.30	.34	.38
13	.09	.43	.75	.01	.22	.39	.53	.65	.76	.86	.94	.02	.09	.15	.21	.27	.31	.36	.40
14	.09	.43	.75	.01	.22	.39	.54	.66	.77	.86	.95	.03	.10	.16	.22	.28	.32	.37	.41
15	0.09	0.43	0.75	1.01	1.22	1.39	1.54	1.66	1.77	1.87	1.95	2.03	2.11	2.17	2.23	2.29	2.34	2.38	2.43
16	.09	.43	.75	.01	.22	.39	.54	.67	.78	.87	.96	.04	.11	.18	.24	.30	.34	.39	.44
17	.09	.43	.75	.01	.22	.40	.55	.67	.78	.88	.97	.05	.12	.19	.25	.30	.35	.40	.45
18	.09	.43	.75	.02	.22	.40	.55	.67	.79	.88	.97	.05	.12	.19	.25	.31	.36	.41	.45
19	.09	.43	.75	.02	.23	.40	.55	.68	.79	.89	.98	.05	.13	.20	.26	.32	.37	.42	.46
20	0.09	0.43	0.75	1.02	1.23	1.40	1.55	1.68	1.79	1.89	1.98	2.06	2.13	2.20	2.27	2.32	2.37	2.42	2.47
24	.09	.43	.75	.02	.23	.41	.56	.69	.80	.90	.99	.08	.15	.22	.28	.34	.39	.45	.49
30	.09	.43	.76	.02	.24	.41	.57	.70	.81	.92	2.01	.09	.17	.24	.30	.36	.41	.47	.52
40	.09	.43	.76	.02	.24	.42	.57	.71	.82	.93	.02	.10	.18	.26	.32	.38	.43	.49	.54
60	0.09	0.43	0.76	1.02	1.24	1.43	1.58	1.72	1.83	1.94	2.04	2.12	2.20	2.28	2.34	2.40	2.46	2.52	2.57
120	.09	.43	.76	.03	.25	.43	.59	.73	.85	.96	.06	.14	.22	.30	.36	.43	.49	.54	.60
∞	0.09	0.43	0.76	1.03	1.25	1.44	1.60	1.74	1.86	1.97	2.07	2.16	2.24	2.32	2.39	2.45	2.52	2.57	2.62

Table 10.4a   (continued)
Upper 5% points

v\n	2	3	4	5	6	7	8	9	10	11	12	13	14	15	16	17	18	19	20
1	18.0	27.0	32.8	37.1	40.4	43.1	45.4	47.4	49.1	50.6	52.0	53.2	54.3	55.4	56.3	57.2	58.0	58.8	59.6
2	6.09	8.3	9.8	10.9	11.7	12.4	13.0	13.5	14.0	14.4	14.7	15.1	15.4	15.7	15.9	16.1	16.4	16.6	16.8
3	4.50	5.91	6.82	7.50	8.04	8.48	8.85	9.18	9.46	9.72	9.95	10.15	10.35	10.52	10.69	10.84	10.98	11.11	11.24
4	3.93	5.04	5.76	6.29	6.71	7.05	7.35	7.60	7.83	8.03	8.21	8.37	8.52	8.66	8.79	8.91	9.03	9.13	9.23
5	3.64	4.60	5.22	5.67	6.03	6.33	6.58	6.80	6.99	7.17	7.32	7.47	7.60	7.72	7.83	7.93	8.03	8.12	8.21
6	3.46	4.34	4.90	5.31	5.63	5.89	6.12	6.32	6.49	6.65	6.79	6.92	7.03	7.14	7.24	7.34	7.43	7.51	7.59
7	3.34	4.16	4.68	5.06	5.36	5.61	5.82	6.00	6.16	6.30	6.43	6.55	6.66	6.76	6.85	6.94	7.02	7.09	7.17
8	3.26	4.04	4.53	4.89	5.17	5.40	5.60	5.77	5.92	6.05	6.18	6.29	6.39	6.48	6.57	6.65	6.73	6.80	6.87
9	3.20	3.95	4.42	4.76	5.02	5.24	5.43	5.60	5.74	5.87	5.98	6.09	6.19	6.28	6.36	6.44	6.51	6.58	6.64
10	3.15	3.88	4.33	4.65	4.91	5.12	5.30	5.46	5.60	5.72	5.83	5.93	6.03	6.11	6.20	6.27	6.34	6.40	6.47
11	3.11	3.82	4.26	4.57	4.82	5.03	5.20	5.35	5.49	5.61	5.71	5.81	5.90	5.99	6.06	6.14	6.20	6.26	6.33
12	3.08	3.77	4.20	4.51	4.75	4.95	5.12	5.27	5.40	5.51	5.62	5.71	5.80	5.88	5.95	6.03	6.09	6.15	6.21
13	3.06	3.73	4.15	4.45	4.69	4.88	5.05	5.19	5.32	5.43	5.53	5.63	5.71	5.79	5.86	5.93	6.00	6.05	6.11
14	3.03	3.70	4.11	4.41	4.64	4.83	4.99	5.13	5.25	5.36	5.46	5.55	5.64	5.72	5.79	5.85	5.92	5.97	6.03
15	3.01	3.67	4.08	4.37	4.60	4.78	4.94	5.08	5.20	5.31	5.40	5.49	5.58	5.65	5.72	5.79	5.86	5.90	5.96
16	3.00	3.65	4.05	4.33	4.56	4.74	4.90	5.03	5.15	5.26	5.35	5.44	5.52	5.59	5.66	5.72	5.79	5.84	5.90
17	2.98	3.63	4.02	4.30	4.52	4.71	4.86	4.99	5.11	5.21	5.31	5.39	5.47	5.55	5.61	5.68	5.74	5.79	5.84
18	2.97	3.61	4.00	4.28	4.49	4.67	4.82	4.96	5.07	5.17	5.27	5.35	5.43	5.50	5.57	5.63	5.69	5.74	5.79
19	2.96	3.59	3.98	4.25	4.47	4.65	4.79	4.92	5.04	5.14	5.23	5.32	5.39	5.46	5.53	5.59	5.65	5.70	5.75
20	2.95	3.58	3.96	4.23	4.45	4.62	4.77	4.90	5.01	5.11	5.20	5.28	5.36	5.43	5.49	5.55	5.61	5.66	5.71
24	2.92	3.53	3.90	4.17	4.37	4.54	4.68	4.81	4.92	5.01	5.10	5.18	5.25	5.32	5.38	5.44	5.50	5.54	5.59
30	2.89	3.49	3.84	4.10	4.30	4.46	4.60	4.72	4.83	4.92	5.00	5.08	5.15	5.21	5.27	5.33	5.38	5.43	5.48
40	2.86	3.44	3.79	4.04	4.23	4.39	4.52	4.63	4.74	4.82	4.91	4.98	5.05	5.11	5.16	5.22	5.27	5.31	5.36
60	2.83	3.40	3.74	3.98	4.16	4.31	4.44	4.55	4.65	4.73	4.81	4.88	4.94	5.00	5.06	5.11	5.16	5.20	5.24
120	2.80	3.36	3.69	3.92	4.10	4.24	4.36	4.48	4.56	4.64	4.72	4.78	4.84	4.90	4.95	5.00	5.05	5.09	5.13
∞	2.77	3.31	3.63	3.86	4.03	4.17	4.29	4.39	4.47	4.55	4.62	4.68	4.74	4.80	4.85	4.89	4.93	4.97	5.01

n is the size of sample from which the range is obtained and v is the number of degrees of freedom of s.
This table is taken from Pearson and Hartley 1956, page 176. By permission.

Table 10.4b   Test Statistics $C3 = \dfrac{X_n - X_i}{S_v}$

Lower 1% points

v\n	2	3	4	5	6	7	8	9	10	11	12	13	14	15	16	17	18	19	20
10	0.02	0.18	0.42	0.64	0.81	0.96	1.11	1.23	1.34	1.41	1.50	1.57	1.62	1.70	1.74	1.81	1.84	1.88	1.92
11	.02	.18	.42	.64	.82	.97	.12	.24	.35	.43	.52	.58	.64	.71	.76	.82	.86	.91	.94
12	.02	.18	.42	.64	.82	.98	.12	.24	.35	.44	.53	.60	.65	.73	.77	.84	.88	.92	.96
13	.02	.18	.42	.64	.83	.98	.13	.25	.36	.45	.54	.61	.66	.74	.79	.85	.89	.94	.98
14	.02	.18	.42	.65	.83	.99	.13	.25	.37	.46	.55	.62	.68	.76	.80	.87	.91	.95	2.00
15	0.02	0.18	0.42	0.65	0.83	0.99	1.14	1.26	1.37	1.46	1.55	1.63	1.69	1.76	1.81	1.88	1.92	1.97	2.01
16	.02	.18	.42	.65	.83	.99	.14	.26	.37	.47	.56	.63	.70	.77	.82	.89	.93	.98	.02
17	.02	.18	.42	.65	.84	1.00	.14	.27	.38	.48	.57	.64	.70	.78	.83	.90	.94	.99	.04
18	.02	.18	.42	.65	.84	.00	.15	.27	.38	.48	.57	.65	.71	.79	.84	.91	.95	2.00	.05
19	.02	.18	.43	.65	.84	.00	.15	.28	.39	.48	.58	.65	.72	.80	.85	.91	.96	.01	.06
20	0.02	0.18	0.43	0.65	0.84	1.01	1.15	1.28	1.39	1.49	1.58	1.66	1.72	1.80	1.85	1.92	1.97	2.01	2.06
24	.02	.18	.43	.65	.85	.01	.16	.29	.40	.50	.60	.67	.74	.82	.88	.94	.99	.05	.09
30	.02	.18	.43	.66	.85	.02	.17	.30	.41	.52	.61	.69	.76	.84	.90	.97	2.02	.07	.12
40	.02	.18	.43	.66	.85	.02	.18	.31	.43	.53	.63	.71	.79	.86	.92	.99	.04	.10	.15
60	0.02	0.18	0.43	0.66	0.86	1.03	1.19	1.32	1.44	1.55	1.64	1.73	1.81	1.88	1.95	2.02	2.07	2.13	2.18
120	.02	.18	.43	.66	.86	.04	.20	.33	.45	.56	.66	.75	.83	.91	.98	.04	.10	.16	.21
∞	0.02	0.19	0.43	0.66	0.87	1.05	1.20	1.34	1.47	1.58	1.68	1.77	1.86	1.93	2.01	2.08	2.14	2.20	2.25

Table 10.4b (continued)

Upper 1% points

v\n	2	3	4	5	6	7	8	9	10	11	12	13	14	15	16	17	18	19	20
1	90.0	135	164	186	202	216	227	237	246	253	260	266	272	277	282	286	290	294	298
2	14.0	19.0	22.3	24.7	26.6	28.2	29.5	30.7	31.7	32.6	33.4	34.1	34.8	35.4	36.0	36.5	37.0	37.5	37.9
3	8.26	10.6	12.2	12.3	14.2	15.0	15.6	16.2	16.7	17.1	17.5	17.9	18.2	18.5	18.8	19.1	19.3	19.5	19.8
4	6.51	8.12	9.17	9.96	10.6	11.1	11.5	11.9	12.3	12.6	12.8	13.1	13.3	13.5	13.7	13.9	14.1	14.2	14.4
5	5.70	6.97	7.80	8.42	8.91	9.32	9.67	9.97	10.24	10.48	10.70	10.89	11.08	11.24	11.40	11.55	11.68	11.81	11.93
6	5.24	6.33	7.03	7.56	7.97	8.32	8.61	8.87	9.10	9.30	9.49	9.65	9.81	9.95	10.08	10.21	10.32	10.43	10.54
7	4.95	5.92	6.54	7.01	7.37	7.68	7.94	8.17	8.37	8.55	8.71	8.86	9.00	9.12	9.24	9.35	9.46	9.55	9.65
8	4.74	5.63	6.20	6.63	6.96	7.24	7.47	7.68	7.87	8.03	8.18	8.31	8.44	8.55	8.66	8.76	8.85	8.94	9.03
9	4.60	5.43	5.96	6.35	6.66	6.91	7.13	7.32	7.49	7.65	7.78	7.91	8.03	8.13	8.23	8.32	8.41	8.49	8.57
10	4.48	5.27	5.77	6.14	6.43	6.67	6.87	7.05	7.21	7.36	7.48	7.60	7.71	7.81	7.91	7.99	8.07	8.15	8.22
11	4.39	5.14	5.62	5.97	6.25	6.48	6.67	6.84	6.99	7.13	7.25	7.36	7.46	7.56	7.65	7.73	7.81	7.88	7.95
12	4.32	5.04	5.50	5.84	6.10	6.32	6.51	6.67	6.81	6.94	7.06	7.17	7.26	7.36	7.44	7.52	7.59	7.66	7.73
13	4.26	4.96	5.40	5.73	5.98	6.19	6.37	6.53	6.67	6.79	6.90	7.01	7.10	7.19	7.27	7.34	7.42	7.48	7.55
14	4.21	4.89	5.32	5.63	5.88	6.08	6.26	6.41	6.54	6.66	6.77	6.87	6.96	7.05	7.12	7.20	7.27	7.33	7.39
15	4.17	4.83	5.25	5.56	5.80	5.99	6.16	6.31	6.44	6.55	6.66	6.76	6.84	6.93	7.00	7.07	7.14	7.20	7.26
16	4.13	4.78	5.19	5.49	5.72	5.92	6.08	6.22	6.35	6.46	6.56	6.66	6.74	6.82	6.90	6.97	7.03	7.09	7.15
17	4.10	4.74	5.14	5.43	5.66	5.85	6.01	6.15	6.27	6.38	6.48	6.57	6.66	6.73	6.80	6.87	6.94	7.00	7.05
18	4.07	4.70	5.09	5.38	5.60	5.79	5.94	6.08	6.20	6.31	6.41	6.50	6.58	6.65	6.72	6.79	6.85	6.91	6.96
19	4.05	4.67	5.05	5.33	5.55	5.73	5.89	6.02	6.14	6.25	6.34	6.43	6.51	6.58	6.65	6.72	6.78	6.84	6.89
20	4.02	4.64	5.02	5.29	5.51	5.69	5.84	5.97	6.09	6.19	6.29	6.37	6.45	6.52	6.59	6.65	6.71	6.76	6.82
24	3.96	4.54	4.91	5.17	5.37	5.54	5.69	5.81	5.92	6.02	6.11	6.19	6.26	6.33	6.39	6.45	6.51	6.58	6.61
30	3.89	4.45	4.80	5.05	5.24	5.40	5.54	5.65	5.76	5.85	5.93	6.01	6.08	6.14	6.20	6.26	6.31	6.36	6.41
40	3.82	4.37	4.70	4.93	5.11	5.27	5.39	5.50	5.60	5.69	5.77	5.84	5.90	5.96	6.02	6.07	6.12	6.17	6.21
60	3.76	4.28	4.60	4.82	4.99	5.13	5.25	5.36	5.45	5.53	5.60	5.67	5.73	5.79	5.84	5.89	5.93	5.98	6.02
120	3.70	4.20	4.50	4.71	4.87	5.01	5.12	5.21	5.30	5.38	5.44	5.51	5.56	5.61	5.66	5.71	5.75	5.79	5.83
∞	3.64	4.12	4.40	4.60	4.76	4.88	4.99	5.08	5.16	5.23	5.29	5.35	5.40	5.45	5.49	5.54	5.57	5.61	5.65

n is the size of the sample from which the range is obtained and v is the number of degrees of freedom of s.

This table is taken from Pearson and Hartley 1956, page 177. By permission.

Table 10.5   Test Statistics $C_1 = \dfrac{(X_n - \bar{X})}{S_v}$

A. Lower percent points of the studentized extreme deviate $(X_n - \bar{X})/S_v$ **or** $(\bar{X} - X_1)/s_v$

	5%							1%						
n →	3	4	5	6	7	8	9	3	4	5	6	7	8	9
v ↓														
10	0.20	0.35	0.46	0.55	0.62	0.69	0.74	0.09	0.19	0.29	0.37	0.43	0.49	0.54
15	0.20	0.35	0.46	0.55	0.63	0.70	0.75	0.09	0.19	0.29	0.37	0.44	0.50	0.56
30	0.20	0.35	0.46	0.56	0.64	0.70	0.77	0.09	0.20	0.29	0.38	0.45	0.51	0.57
∞	0.20	0.35	0.47	0.56	0.65	0.72	0.78	0.09	0.20	0.30	0.38	0.46	0.53	0.59

B. Upper per cent points of the studentized extreme deviate

	5%							1%						
n →	3	4	5	6	7	8	9	3	4	5	6	7	8	9
v ↓														
10	2.02	2.29	2.49	2.63	2.75	2.85	2.93	2.76	3.05	3.25	3.39	3.50	3.59	3.67
11	1.99	2.26	2.44	2.58	2.70	2.79	2.87	2.71	3.00	3.19	3.33	3.44	3.53	3.61
12	1.97	2.22	2.40	2.54	2.65	2.75	2.83	2.67	2.95	3.14	3.28	3.39	3.48	3.55
13	1.95	2.20	2.38	2.51	2.62	2.71	2.79	2.63	2.91	3.10	3.24	3.34	3.43	3.51
14	1.93	2.18	2.35	2.48	2.59	2.68	2.70	2.60	2.87	3.06	3.20	3.30	3.39	3.47
15	1.92	2.16	2.33	2.46	2.56	2.65	2.73	2.57	2.84	3.02	3.16	3.27	3.35	3.43
16	1.90	2.14	2.31	2.44	2.54	2.63	2.70	2.55	2.81	3.00	3.13	3.24	3.32	3.39
17	1.89	2.13	2.30	2.42	2.52	2.61	2.68	2.52	2.79	2.97	3.10	3.21	3.29	3.36
18	1.88	2.12	2.28	2.41	2.51	2.59	2.66	2.50	2.77	2.95	3.08	3.18	3.27	3.34
19	1.87	2.11	2.27	2.39	2.49	2.58	2.65	2.49	2.75	2.92	3.06	3.10	3.24	3.31
20	1.87	2.10	2.26	2.38	2.48	2.56	2.63	2.47	2.73	2.91	3.04	3.14	3.22	3.29
24	1.84	2.07	2.23	2.35	2.44	2.52	2.59	2.43	2.68	2.85	2.97	3.07	3.15	3.22
30	1.82	2.04	2.20	2.31	2.40	2.48	2.55	2.38	2.62	2.79	2.91	3.01	3.08	3.15
40	1.80	2.02	2.17	2.28	2.37	2.44	2.51	2.34	2.57	2.73	2.85	2.94	3.02	3.08
60	1.78	1.99	2.14	2.25	2.33	2.41	2.47	2.30	2.52	2.68	2.79	2.88	2.95	3.01
120	1.76	1.97	2.11	2.21	2.30	2.37	2.43	2.25	2.48	2.62	2.73	2.82	2.89	2.95
∞	1.74	1.94	2.08	2.18	2.27	2.33	2.39	2.22	2.43	2.57	2.68	2.70	2.83	2.88

V is degrees of freedom of S_v

This table is taken from the paper of Nair 1948, page 143. By permission. Biometrika, Vol. 35.

Twenty similar measurements give an S_v of 2.66 mg/m^3. Calculate the mean $\overline{X}$ to get $\overline{X}$ = 9.43.

Then use test $C_1 = \dfrac{(X_n - \overline{X})}{S_v}$ or $\dfrac{(\overline{X} - X_{1)}}{S_v}$

The question, then, to be answered is: Should the apparently high value of 15.33 be rejected?

To solve the problem, use test C_1.

The mean $\overline{X}$ = 9.43

$$S_v = 2.66 \text{ (given)}$$

$$C_1 = \frac{15.33 - 9.43}{2.66} = \frac{5.9}{2.66} = 2.22$$

The ratio would have to be greater than 2.65 for n = 10 and d.f. = 19, at the 5% confidence level from Table 10.5. Since the independent measure of S is based on 20 measurements, the d.f. figure used is one less than 20, or 19. Using Table 10.6, we obtain a critical value (CV) of 2.70 for N = 10 and d.f. = 19 at the 5% level.

The statistic calculated, 2.22, is smaller than either of the table values, so the result of 15.33 is not rejected.

Example 10.5

Ten total dust measurements were made. The calculated values of the results were:

TOTAL DUST CONCENTRATIONS MG/M$_3$

11.57	8.36	9.06	11.29	9.27
7.32	11.23	9.37	10.56	10.00

Similar measurements, taken at a different time, yield an S_v of 0.89, with 19 d.f. Now calculate the mean $\overline{X}$:

Use the test $C_1 = \dfrac{(X_n - \overline{X})}{S_v}$ or $\dfrac{(\overline{X} - X_{1})}{S_v}$

Determine whether the high or low value should be rejected.

Use test C_1 from Tables 10.5 and 10.6

Table 10.6   Test Statistic $C1 = \dfrac{(X_n - \overline{X})}{S_v}$

Critical Values for T When Standard Deviation s, is Independent of Present Sample

n $v = df$	3	4	5	6	7	8	9	10	12
					1% points				
10	2.78	3.10	3.32	3.48	3.62	3.73	3.82	3.90	4.04
11	2.72	3.02	3.24	3.39	3.52	3.63	3.72	3.79	3.93
12	2.67	2.96	3.17	3.32	3.45	3.55	3.64	3.71	3.84
13	2.63	2.92	3.12	3.27	3.38	3.48	3.57	3.64	3.76
14	2.60	2.88	3.07	3.22	3.33	3.43	3.51	3.58	3.70
15	2.57	2.84	3.03	3.17	3.29	3.38	3.46	3.53	3.65
16	2.54	2.81	3.00	3.14	3.25	3.34	3.42	3.49	3.60
17	2.52	2.79	2.97	3.11	3.22	3.31	3.38	3.45	3.56
18	2.50	2.77	2.95	3.08	3.19	3.28	3.35	3.42	3.53
19	2.49	2.75	2.93	3.06	3.16	3.25	3.33	3.39	3.50
20	2.47	2.73	2.91	3.04	3.14	3.23	3.30	3.37	3.47
24	2.42	2.68	2.84	2.97	3.07	3.16	3.23	3.29	3.38
30	2.38	2.62	2.79	2.91	3.01	3.08	3.15	3.21	3.30
40	2.34	2.57	2.73	2.85	2.94	3.02	3.08	3.13	3.22
60	2.29	2.52	2.68	2.79	2.88	2.95	3.01	3.06	3.15
120	2.25	2.48	2.62	2.73	2.82	2.89	2.95	3.00	3.08
∞	2.22	2.43	2.57	2.68	2.76	2.83	2.88	2.93	3.01
					5% points				
10	2.01	2.27	2.46	2.60	2.72	2.81	2.89	2.96	3.08
11	1.98	2.24	2.42	2.56	2.67	2.76	2.84	2.91	3.03
12	1.96	2.21	2.39	2.52	2.63	2.72	2.80	2.87	2.98
13	1.94	2.19	2.36	2.50	2.60	2.69	2.76	2.83	2.94
14	1.93	2.17	2.34	2.47	2.57	2.66	2.74	2.80	2.91
15	1.91	2.15	2.32	2.45	2.55	2.64	2.71	2.77	2.88
16	1.90	2.14	2.31	2.43	2.53	2.62	2.69	2.75	2.86
17	1.89	2.13	2.29	2.42	2.52	2.60	2.67	2.73	2.84
18	1.88	2.11	2.28	2.40	2.50	2.58	2.65	2.71	2.82
19	1.87	2.11	2.27	2.39	2.49	2.57	2.64	2.70	2.80
20	1.87	2.10	2.26	2.38	2.47	2.56	2.63	2.68	2.78
24	1.84	2.07	2.23	2.34	2.44	2.52	2.58	2.64	2.74
30	1.82	2.04	2.20	2.31	2.40	2.48	2.54	2.60	2.69
40	1.80	2.02	2.17	2.28	2.37	2.44	2.50	2.56	2.65
60	1.78	1.99	2.14	2.25	2.33	2.41	2.47	2.52	2.61
120	1.76	1.96	2.11	2.22	2.30	2.37	2.43	2.48	2.57
∞	1.74	1.94	2.08	2.18	2.27	2.33	2.39	2.44	2.52

The above percentage points are reproduced from H. A. David, "Revised upper percentage points of the extreme studentized deviate from the sample mean," Biometrika, Vol. 43 (1956), pp. 449-451. (Reference [3]). This table is taken from Grubbs 1969, page 15.

From Table 10.5	2.65	3.31
	U5%	U1%
From Table 10.6	2.70	3.39
	U5%	U1%

Therefore, since the calculated value 1.98 is less than, and the calculated value of 2.78 is greater than, the table critical values at the 95% level, retain the high value of 11.57 and reject the low value of 7.32.

BOTH μ AND σ UNKNOWN

This situation is one often encountered by the laboratory worker. The only information at hand is from the set of observations themselves. The array of tests available parallels that of the previous cases, but, in addition, Dixon (1953) has provided a series of range and sub-range tests that can quickly and easily deal with suspected outlier situations.

The efficiency of these tests is similar to that of earlier cases. Test statistic D_6 was designed to deal with two outliers, both lying to the right of the data group, and will be discussed in the next chapter.

The Dixon series of tests are one-sided. David (1970) concludes that the D_1 test is best for one-sided tests, when the number of observations (n) increase. When the number of observations (n) is small, D_1, D_4, and r_{10} are equivalent. For the two-sided tests, D_2 and D_5 are about equal in performance. Where there are two suspected outliers in a single set of data, test statistic D_5 outperforms D_2. The Dixon ratio tests are not designed to handle this situation at all. Grubbs (1979) Confirms the value of the D_1 test for a single outlier. The series of Dixon ratio tests follow:

	If X_n is Suspect	If X_1 is Suspect
$r_{10} =$ n = 3 to 7	$\dfrac{X_n\, X_{n-1}}{X_n - X_1}$	$\dfrac{X_2 - X_1}{X_n - X_1}$
$r_{11} =$ n = 8 to 10	$\dfrac{X_n - X_{n-1}}{X_n - X_2}$	$\dfrac{X_2 - X_1}{X_{n-1} - X_1}$
$r_{21} =$ n = 11-13	$\dfrac{X_n - X_{n-2}}{X_n - X_2}$	$\dfrac{X_3 - X_1}{X_{n-1} - X_1}$
$r_{22} =$ n = 14 to 25	$\dfrac{X_n - X_{n-2}}{X_n - X_3}$	$\dfrac{X_3 - X_1}{X_{n-2} - X_i}$

Two examples of how to use the test statistic D_1 follow.

Example 10.6.

This example will use the data from Example 10.1, without having prior knowledge of the population mean or standard deviation.

$$\text{We will use the test statistic } D_1 = \frac{X_n - \overline{X}}{s}$$

1. Calculate the mean, $\overline{X}$.
2. Calculate the standard deviation, s.
3. Determine whether any of the data points should be rejected.
1. $\overline{X} = 9.82$
2. $s = 2.66$

$$3. \quad D_1 = \frac{X_n - \overline{X}}{s} = \frac{14.6 - 9.82}{2.66} = \frac{4.78}{2.66} = 1.80$$

Using Table 10.7, with n = 20, we find:

Test	$CV_{5\%}$	$CV_{2.5\%}$	$CV_{1\%}$
1.80	2.56	2.71	2.88

Value
Since the calculated test value is less than the table critical values at the 5%, 2.5%, and 1.0% confidence levels, we do not reject any data points.

Example 10.7

This example will use the data from Example 10.2, without having prior knowledge of the population mean or standard deviation.

$$\text{We will again use the test statistic } D_1 = \frac{X_n - \overline{X}}{s}$$

1. Calculate the mean $\overline{X}$.
2. Calculate the standard deviation s.
3. Determine whether any of the data points should be rejected.

Table 10.7   Test Statistic $D1 = \dfrac{X_n - \overline{X}}{s}$ or $\dfrac{\overline{X} - X_i}{s}$

Number of Observations n	5% Significance Level	2.5% Significance Level	1% Significance Level
3	1.15	1.15	1.15
4	1.46	1.48	1.49
5	1.67	1.71	1.75
6	1.82	1.89	1.94
7	1.94	2.02	2.10
8	2.03	2.13	2.22
9	2.11	2.21	2.32
10	2.18	2.29	2.41
11	2.23	2.36	2.48
12	2.29	2.41	2.55
13	2.33	2.46	2.61
14	2.37	2.51	2.66
15	2.41	2.55	2.71
16	2.44	2.59	2.75
18	2.50	2.65	2.82
19	2.53	2.68	2.85
20	2.56	2.71	2.88
21	2.58	2.73	2.91
22	2.60	2.76	2.94
23	2.62	2.78	2.96
24	2.64	2.80	2.99
25	2.66	2.82	3.01
30	2.75	2.91	
35	2.82	2.98	
40	2.87	3.04	
45	2.92	3.09	
50	2.96	3.13	
60	3.03	3.20	
70	3.09	3.26	
80	3.14	3.31	
90	3.18	3.35	
100	3.21	3.38	

$$D_{1_n} = \frac{x_n - \overline{x}}{s} \qquad s = \left\{ \frac{\Sigma(x_i - \overline{x})^2}{n - 1} \right\}^{\frac{1}{2}} = \left\{ \frac{n \, \Sigma \, x^2 - (\Sigma \, x_i)^2}{n(n - 1)} \right\}^{\frac{1}{2}}$$

$$D_{1_n} = \frac{\overline{x} - x_1}{s} \qquad x_i \le x_2 \le \dots \le x_n$$

Note: Values of D for n $\le$ 25 are based on those given in Reference [8]. For n > 25, the values of D are approximated. All values have been adjusted for division by n - 1 instead of n in calculating s.

This table is taken from Grubbs 1969, page 4.

Table 10.8 Test Statistic $D_3 = \dfrac{\text{Range }(w)}{s}$ or $\dfrac{X_n - X_1}{s}$

Critical Values for w/s (Ratio of Range to Sample Standard Deviation)*

Number of Observations n	5% Significance Level	1% Significance Level	0.5% Significance Level
3	2.00	2.00	2.00
4	2.43	2.44	2.45
5	2.75	2.80	2.81
6	3.01	3.10	3.12
7	3.22	3.34	3.37
8	3.40	3.54	3.58
9	3.55	3.72	3.77
10	3.68	3.88	3.94
11	3.80	4.01	4.08
12	3.91	4.13	4.21
13	4.00	4.24	4.32
14	4.09	4.34	4.43
15	4.17	4.43	4.53
16	4.24	4.51	4.62
17	4.31	4.59	4.69
18	4.38	4.66	4.77
19	4.43	4.73	4.84
20	4.49	4.79	4.91
30	4.89	5.25	5.39
40	5.15	5.54	5.69
50	5.35	5.77	5.91
60	5.50	5.93	6.09
80	5.73	6.18	6.35
100	5.90	6.36	6.54
150	6.18	6.64	6.84
200	6.38	6.85	7.03
500	6.94	7.42	7.60
1000	7.33	7.80	7.99

*Taken from H. A. David, H. O. Hartley and E. S. Pearson, "The Distribution of the Ratio in a Single Sample of Range to Standard Deviation," Biometrika, Vol. 41 (1954), pp. 482-493. (Reference [4])

$w = x_n - x_1$ $s = \sqrt{\dfrac{\Sigma (x_i - x)^2}{n - 1}}$

$x_1 \leq x_2 \leq \ldots \leq x_n$

This table is taken from Grubbs 1969, page 8.

Table 10.9 Dixon Ratio Test Statistics   CRITICAL VALUES AND CRITERIA FOR TESTING FOR EXTREME VALUES

N \ α	.30	.20	.10	.05	.02	.01	.005	Criterion
3	.684	.781	.886	.941	.976	.988	.994	
4	.471	.560	.679	.765	.846	.889	.926	
5	.373	.451	.557	.642	.729	.780	.821	$r_{10} = \dfrac{x_N - x_{N-1}}{x_N - x_1}$
6	.318	.386	.482	.560	.644	.698	.740	
7	.281	.344	.434	.507	.586	.637	.680	
8	.318	.385	.479	.554	.631	.683	.725	
9	.288	.352	.441	.512	.587	.635	.677	$r_{11} = \dfrac{x_N - x_{N-1}}{x_N - x_2}$
10	.265	.325	.409	.477	.551	.597	.639	
11	.391	.442	.517	.576	.638	.679	.713	
12	.370	.419	.490	.546	.605	.642	.675	$r_{21} = \dfrac{x_N - x_{N-2}}{x_N - x_2}$
13	.351	.399	.467	.521	.578	.615	.649	
14	.370	.421	.492	.546	.602	.641	.674	
15	.353	.402	.472	.525	.579	.616	.647	
16	.338	.386	.454	.507	.559	.595	.624	
17	.325	.373	.438	.490	.542	.577	.605	
18	.314	.361	.424	.475	.527	.561	.589	$r_{22} = \dfrac{x_N - x_{N-2}}{x_N - x_3}$
19	.304	.350	.412	.462	.514	.547	.575	
20	.295	.340	.401	.450	.502	.535	.562	
21	.287	.331	.391	.440	.491	.524	.551	
22	.280	.323	.382	.430	.481	.514	.541	
23	.274	.316	.374	.421	.472	.505	.532	
24	.268	.310	.367	.413	.464	.497	.524	
25	.262	.304	.360	.406	.457	.489	.516	

Reproduced from: W. J. Dixon, "Processing Data for Outliers." Biometrics 9:74-89, 1953. With permission of the Biometric Society.

Again, use test D_1 with $\overline{X} = 45.5$ and s = 11.1 Then

$$D_1 = \frac{\overline{X} - X_1}{s} = \frac{45.5 - 21}{11.1} = 2.21$$

Using Table 10.7, with n = 10, we find:

Test	$CV_{5\%}$	$CV_{2.5\%}$	$CV_{1\%}$
2.21	2.18	2.29	2.41

Value

Since the calculated test value 2.21, is greater than the critical value at the 5% confidence level, we do reject the data point. Repeating the test a second time also results in rejection of the suspected outlier.

References

David, H. A. 1970. *Order Statistics*. New York: John Wiley and Sons, Inc.

Dixon, W. J. 1950. Analysis of Extreme Values. *Annals of Mathematical Statistics 21*, pp. 488-506.

Dixon, W. J. 1953. Processing data for outliers. *Biometrics* 9:74-89.

Grubbs, F. E. 1969. Procedures for detecting outlying observations in samples. Technometrics 11:1-21.

Grubbs, F. E. 1979. Procedures for detecting outlying observations. In *Army Statistics Manual DARCOM—P706-103*, Chapter 3. U.S. Army Research and Development Center, Aberdeen Proving Ground, MD 21005.

Irwin, J. O. 1925. On a criterion for the rejection of outlying observations. *Biometrika* 17:238-50.

McKay, A. T. 1935. The distribution of the difference between the extreme observation and the sample mean in samples of n from a normal universe. *Biometrika* 27:466-71.

Nair, K. R. 1948. The distribution of the extreme deviate from the sample mean and its studentized form. *Biometrika* 35:118-44.

Pearson, E. S., and H. O. Hartley. 1956. *Biometrika Tables for Statisticians*. Cambridge, England: Cambridge Press.

"Student." 1927. Errors of routine analysis. *Biometrika 1* 19:151-64.

11

Multiple Outliers

The two previous chapters treated the general problems of outlying observations and the methods of dealing with only one outlier per sample. The approaches to dealing with a single outlier were modified, depending on the nature of and the availability of independent knowledge, about the population variance σ^2. In dealing with multiple outliers, most applications deal with the situation of having no prior knowledge of μ or σ—the "D" type statistic.

This chapter will first deal with the case of handling one high and one low outlier; then, two high or two low outliers; multiple outliers; and finally, a special nonparametric test, the slippage type test.

CASE 1—ONE HIGH AND ONE LOW OUTLIER

The range type test is recommended in this case because it is designed to handle upper and lower outliers, in contrast to the case of the outliers being only on one side of the distribution. The test statistic of David, Hartley, and Pearson (1966) has been chosen for ASTM E-11 use, by Grubbs (1969).

$$\text{The test is: } R = \frac{w}{s} \tag{11-1}$$

$$\text{Where } w = X_n - X_1 \tag{11-2}$$

$$s = \text{the sample standard deviation}$$

$$s = \sqrt{\frac{\Sigma(X_1 - \overline{X})^2}{n - 1}}$$

(11-3)

The classic case to which, it seems, all tests are applied is the set of 15 observations of the "vertical semidiameters of Venus," made by a Lieutenant Herndon in 1846 and discussed by Chauvenet in 1876. The residuals in X_1, X_2, and so on, order are:

-1.40	-0.24	-0.05	0.18	0.48
-0.44	-0.22	0.06	0.20	0.63
-0.33	-0.13	0.10	0.39	1.01

Using the data set above, with 1.01 being the apparent outlier on the high side and -1.40 on the low side of the data package, we can calculate that w = 1.01 - (-1.40) = 2.41 and s = 0.551 as $\overline{X}$ = 0.016. The test statistic or ratio is:

$$R = \frac{w}{s} = \frac{2.41}{0.553} = 4.36$$

(11-4)

The Table of Critical Values for this test (see Table 10.8) (the same table we used for test D_3) is entered for n = 15. The test statistic value of 4.36 falls between the critical values for the 5% (4.17) and the 1% (4.43) confidence levels. If both the X_1 (-1.40) and the X (1.01) values were about equidistant from the mean, one could conclude that both values were outliers. Since, however, this is not a true t-statistic type test, such as the D_1, test:

$$D_1 = \frac{X_n - \overline{X}}{s} \text{ or } \frac{\overline{X} - X_1}{s}, \text{ the test}$$

(11-5)

should be used to test the largest suspect outlier observation -1.40. Applying the test D_1, reject the value -1.40. A second application of the D_1 test to the other suspect outlier, 1.01, against the new total of 14 observations fails to reject the value. The Dixon ratio test could also be applied to these confirmatory tests as well, and will yield equivalent decisions.

TWO HIGH OR TWO LOW OUTLIERS

This case has been well documented by Grubbs (1950 and 1969), and Grubbs and Beck (1972). MacMillan and David (1971) have also studied such a case, when the variance is known, and MacMillan (1971) and MacMillan and Moran (1973), when the variance is unknown. The Grubbs test was mentioned in the last chapter as test statistic D_6.

The test statistic D_6 is abbreviated as: $D_6 = \dfrac{S^2_{1.2}}{S^2}$

(11-6)

$$\frac{S^2_{n-1.n}}{S^2} \text{ or } \frac{S^2_{1.2}}{S_2} \text{ where } S^2 = \sum_{i-1}^{n} (X_i - \overline{X})^2$$

(11-7)

$$\text{where } \overline{X} = \frac{1}{n} \sum_{i=1}^{n} X_i, S^2_{n-1.n} = \sum_{i=1}^{n-2} \frac{(X_i - \overline{X}_{n-1.n})^2}{n-2}$$

(11-8)

$$\text{where } \overline{X}_n - 1, n = \frac{1}{n-2} \sum_{i=1}^{n-2} X_i \, S^2_{1.2} = \sum_{i=3}^{n}$$

(11-9)

$$\frac{(X_i - \overline{X}_{1.2})^2}{n-2} \text{ where } \overline{X}_{1.2} = \frac{1}{n-2} \sum_{i=3}^{n} X_i$$

(11-10)

Table 11.1 gives the critical values for the test statistic D_6. The example set of data for this test consists of ten observations. Put in ascending order, they are:

2.02	2.22	3.04	3.23	3.59
3.73	3.94	4.05	4.11	4.13

Applying the test statistic $\dfrac{S_{1.2}}{S^2}$ would determine whether the values 2.02 and 2.22 should be rejected. Using the formulas above, we calculate:

$$S^2 = \sum_{i=1}^{n} (X_i - \overline{X})^2 = \frac{n \sum X_i^2 - (\sum X_i)^2}{n}$$

(11-11)

Table 11.1   Critical Values for $S^2_{n-1,n}/S^2$ or $S^2_{1,2}/S^2$ for Simultaneously Testing the Two Largest or Two Smallest Observations*

Number of Observations n	10% Significance Level	5% Significance Level	1% Significance Level
4	.0031	.0008	.0000
5	.0376	.0183	.0035
6	.0921	.0565	.0186
7	.1479	.1020	.0440
8	.1994	.1478	.0750
9	.2454	.1909	.1082
10	.2853	.2305	.1415
11	.3226	.2666	.1736
12	.3552	.2996	.2044
13	.3843	.3295	.2333
14	.4106	.3568	.2605
15	.4345	.3818	.2859
16	.4562	.4048	.3098
17	.4761	.4259	.3321
18	.4944	.4455	.3530
19	.5113	.4636	.3725
20	.5269	.4804	.3909

*These significance levels are taken from Table V of Grubbs, Reference [8]. An observed ratio less than the appropriate critical ratio in this table calls for rejection of the null hypothesis.

This Table was taken from Grubbs 1969 page 11. By permission. Chapter 3, Army Statistics Manual, DARCOM p. 706-1-3.

$$= \frac{10(121.3594) - (34.06)^2}{10} \quad \text{or} \quad S^2 = 5.351$$

$$S^2_{1.2} = \sum_{i=3}^{n} (X_i - X_{1.2})^2 \tag{11-12}$$

$$= \frac{(n-2) \sum_{i=1}^{n} X_i^2 - (\sum_{i=3}^{n} X_i)^2}{n-2}$$

$$= \frac{8(112.3506) - (29.82)^2}{8} = \frac{9.5724}{8} = 1.197$$

$$D_6 = \frac{S^2_{1.2}}{S^2} = \frac{1.197}{5.351} = 0.224 \tag{11-13}$$

Comparing the D_6 = 0.224 to the values in Table 11.1 for n = 10, the 10% critical value (CV) is 0.2853, the 5% CV is 0.2305, and the 1% CV is 0.1415. The test value D_6 = 0.224 falls between the 5% and the 1% CVs. We would then reject the null hypothesis of no difference and say that the two low values 2.02 and 2.22 are not from the same population as the rest of this set of data.

It would also be advantageous to apply the appropriate Dixon ratio test to the inside value (either X_2 or X_{n-1})—in this case, 3.04—of the two suspected outliers. If this observation is rejected, then the outside observations would, of course, be considered to be even more abnormal.

REJECTING MULTIPLE OUTLIERS

The tests for rejecting multiple outliers are theoretically more complex, and, therefore, in practice, more complicated than for a single outlier or a one-sided two-observation rejection test. Several of these tests have been developed in recent years. Tietzen and Moore (1972) developed their test for two or more outliers, where some are larger and some are smaller than the remaining values of the sample. This test would have difficulty, however, in properly identifying the outliers of a set of ten observations, where the n-1 value is 10, and the n value is 100 and $\overline{X}$ is 11.

The Rosner (1975) test provides a better solution for such a set of data. The test is set up to detect as many as k outliers. The value of k can be selected as one desires. This chosen value of k is used to trim the set of data values. That is, the k largest and k smallest values in the ranked set are omitted from the sample so that no outliers remain, only "good" values.

The test is then conducted by calculating a "trimmed mean" (a) and a trimmed variance (b^2) as follows:

$$a = \sum_{i = k + 1}^{n - k} \frac{X_i}{n - 2k} \tag{11-14}$$

$$b^2 = \sum_{i = k + 1}^{n - k} \frac{(X_i - a)^2}{n - 2k - 1} \tag{11-15}$$

Rosner uses these values to test the residual R, rather than the original X and s of the whole sample.

$$R_1 = \frac{\max |X_1 - a|}{b}$$

This is compared to the critical value in Table 11-2. If the largest difference value

TABLE 11.2   Percentage Points of Rosner's RST Many outlier Test Statistics

n	α =	R_1 AND R_2 k = 2 0.10	0.05	0.01
10	R_1	7.35 ± 1.02	8.90 ± .146	13.38 ± .748
	R_2	4.92 ± .067	5.92 ± .103	9.13 ± .407
15	R_1	5.28 ± .063	6.01 ± .056	8.10 ± .208
	R_2	3.84 ± .045	4.31 ± .060	5.39 ± .134
20	R_1	4.64 ± .043	5.18 ± .053	6.47 ± .182
	R_2	3.50 ± .024	3.81 ± .032	4.70 ± .095
30	R_1	4.26 ± .027	4.62 ± .037	5.51 ± .108
	R_2	3.31 ± .021	3.57 ± .017	4.15 ± .053
40	R_1	4.04 ± .019	4.41 ± .033	5.26 ± .047
	R_2	3.23 ± .017	3.43 ± .030	3.92 ± .042
50	R_1	3.98 ± .013	4.25 ± .019	4.98 ± .081
	R_2	3.20 ± .011	3.39 ± .022	3.80 ± .047
75	R_1	3.89 ± .016	4.16 ± .016	4.77 ± .074
	R_2	3.19 ± .013	3.37 ± .029	3.72 ± .038
100	R_1	3.83 ± .016	4.09 ± .027	4.66 ± .088
	R_2	3.20 ± .012	3.34 ± .0076	3.74 ± .037

n	α =	R_1, R_2 AND R_3 k = 3 0.10	0.05	0.01
20	R_1	5.91 ± .059	6.60 ± .079	8.19 ± .137
	R_2	4.50 ± .047	5.06 ± 0.52	6.34 ± .151
	R_3	3.73 ± .037	4.16 ± .046	5.22 ± .098
30	R_1	5.07 ± .037	5.60 ± .063	6.88 ± .093
	R_2	3.93 ± .028	4.32 ± .037	5.09 ± .121
	R_3	3.35 ± .016	3.62 ± .039	4.27 ± .076
40	R_1	4.60 ± .037	5.06 ± .040	6.05 ± .103
	R_2	3.68 ± .021	3.92 ± .021	4.53 ± .051
	R_3	3.20 ± .016	3.41 ± .024	3.82 ± .063
50	R_1	4.43 ± .033	4.76 ± .049	5.68 ± .038
	R_2	3.60 ± .014	3.82 ± .018	4.55 ± .086
	R_3	3.14 ± .019	3.30 ± .014	3.77 ± .047
75	R_1	4.18 ± .024	4.46 ± .034	5.10 ± .036
	R_2	3.47 ± .013	3.67 ± .019	4.10 ± .040
	R_3	3.08 ± .0096	3.19 ± .012	3.57 ± .045
100	R_1	4.12 ± .019	4.37 ± .034	4.98 ± .120
	R_2	3.44 ± .012	3.60 ± .022	3.88 ± .039
	R_3	3.10 ± .012	3.21 ± .016	3.45 ± .031

n	α =	R_1, R_2, R_3 AND R_4 k = 4 0.10	0.05	0.01
20	R_1	7.56 ± .083	8.52 ± .112	11.70 ± .340
	R_2	5.88 ± .042	6.53 ± .050	8.83 ± .263
	R_3	4.91 ± .038	5.46 ± .064	7.23 ± .199
	R_4	4.17 ± .035	4.65 ± .056	6.03 ± .116
30	R_1	5.90 ± .030	6.40 ± .055	7.65 ± .096
	R_2	4.63 ± .030	5.01 ± .034	5.90 ± .094
	R_3	3.95 ± .037	4.27 ± .049	5.09 ± .089
	R_4	3.50 ± .024	3.76 ± .034	4.53 ± .101
40	R_1	5.23 ± .036	5.67 ± .066	6.85 ± .264
	R_2	4.13 ± .025	4.47 ± .037	5.24 ± .087
	R_3	3.60 ± .031	3.82 ± .030	4.52 ± .079
	R_4	3.25 ± .020	3.43 ± .027	3.99 ± .043

TABLE 11.2   *(continued)*

n	α =	R_1, R_2, R_3 AND R_4 k = 4		
		0.10	0.05	0.01
50	R_1	4.85 ± .036	5.19 ± .063	6.18 ± .111
	R_2	3.95 ± .022	4.18 ± .028	4.86 ± .082
	R_3	3.46 ± .014	3.67 ± .019	4.20 ± .066
	R_4	3.14 ± .0098	3.30 ± .021	3.75 ± .041
75	R_1	4.55 ± .039	4.87 ± .060	5.66 ± .105
	R_2	3.73 ± .022	3.94 ± .018	4.41 ± .054
	R_3	3.31 ± .010	3.47 ± .020	3.81 ± .021
	R_4	3.04 ± .014	3.16 ± .019	3.50 ± .034
100	R_1	4.43 ± .037	4.67 ± .034	5.38 ± .091
	R_2	3.64 ± .016	3.80 ± .018	4.28 ± .056
	R_3	3.27 ± .012	3.39 ± .011	3.72 ± .037
	R_4	3.03 ± .011	3.14 ± 0.12	3.41 ± .028

(The ± values are standard errors.)

This table is taken from Grubbs 1979, pages 3-65-67. By permission Grubbs table adapted from Rosner 1977.

is rejected, the second largest value is tested against the critical value in the same manner. The Rosner sequential test is particularly powerful against outliers on one side, rather than on both sides. It is flexible in that one does not have to specify an exact number of suspect outliers to be tested, as was necessary in earlier tests.

An example application will clarify how this test is used. We will use the Venus semidiameter values cited earlier, since this data is familiar to us. The values ranked in increasing order of magnitude are:

-1.40	-0.44	-0.30	-0.24	-0.22
-0.13	-0.05	0.06	0.10	0.18
0.20	0.39	0.48	0.63	1.01

1. Since the values -1.40 and 1.01 are suspect, set k as equal to 2 (k = 2).
2. Discard the K (2) lowest values -1.40 and -0.44, and the k (2) highest values 0.63 and 1.01.
3. Calculate the trimmed mean and standard deviation:

$$a = \sum_{i = k + 1}^{n - k} \frac{Xi}{n - 2k} = 0.04273$$

$$b = \left(\sum_{i = k + 1}^{n - k} \frac{(Xi - a)^2}{(n - 2k - 1)} \right)^{1/2} = 0.25$$

4. Calculate R1 and R2:

$$R1 = \frac{\max |X_{i - a}|}{b} = \frac{-1.40 - 0.0427}{0.25} = 5.60$$

$$R2^* = \frac{\max |X_i - a|}{b} = \frac{+1.01 - 0.0427}{0.25} = 3.76$$

*Discarding X_1 used in R_1 value.

The critical values from table:

For n = 15 α =	0.10	0.05	0.01
R1	5.28	6.01	8.10
R2	3.84	4.31	5.39

5. Conclude that the -1.40 value cannot be rejected, even at the 5% level, and the 1.01 value, cannot be rejected not even at the 10% level. These different conclusions seem to point out the weakness of the Rosner test, when the outliers are on both sides.

Hawkins (1978) has an effective multiple outlier test of the same type as the Grubbs test did earlier. The Test statistic E_k equals the inlier sums of squares S_k^{2*} divided by the entire sample sums of squares s^2:

$$E_k^* = \frac{S_k^{2*}}{S^2}$$

The critical values are given in Table 11.3.

For an example problem, illustrating how to use the Hawkins E_k test, we will again use the Venus semidiameter data.

1. Calculate the inlier sums of squares (not including the -1.40 and 1.01 values).

$$s_k^{2*} = .1936 + .1089 + .0576 + .0484 + .0169 + .0025 + .0036 + .01 + .0324 + .04 + .1521 + .2304 + .3969 = 1.29$$

2. Calculate the sums of squares for the entire sample.

$$s^2 = 1.96 + .1936 + .1089 + .0576 + .0484 + .0169 + .0025 + .0036 + .01 + .0324 + .04 + .1521 + .2304 + .3969 + 1.0201.$$
$$s_2 = 4.27$$

3. Calculate the Hawkins test statistic:

$$E_k^* = \frac{S_k^2}{S^2} = \frac{1.29}{4.27} = 0.302$$

4. Conclude that both values should be rejected, since the critical value from Table 11.3 for k = 2 and n = 15, at the 0.05 level, is 0.3104, and the critical value at the 0.01 level is 0.2319. The calculated value of 0.302 falls between the two critical values, so the likelihood of the questionable values being this low and this high due to chance alone is considered to be negligible.

Table 11.3   Percentage Points of Hawkin's E*

d.f. k	n/α	0 — 0.05	0 — 0.01	0 — 0.001	10 — 0.05	10 — 0.01	10 — 0.001	20 — 0.05	20 — 0.01	20 — 0.001
2	5	0.0061	0.0011	0.0001	0.4556	0.3380	0.2205	0.6613	0.5610	0.4433
2	10	0.1640	0.0995	0.0487	0.4920	0.3985	0.2948	0.6461	0.5650	0.4664
2	15	0.3104	0.2319	0.1529	0.5404	0.4592	0.3638	0.6596	0.5890	0.5010
2	20	0.4136	0.3367	0.2509	0.5829	0.5105	0.4222	0.6780	0.6148	0.5345
2	25	0.4886	0.4168	0.3320	0.6186	0.5531	0.4712	0.6964	0.6388	0.5647
2	30	0.5455	0.4792	0.3982	0.6486	0.5887	0.5126	0.7135	0.6606	0.5917
2	40	0.6262	0.5698	0.4977	0.6959	0.6449	0.5782	0.7432	0.6975	0.6369
2	50	0.6810	0.6322	0.5684	0.7314	0.6869	0.6278	0.7676	0.7272	0.6731
2	75	0.7639	0.7277	0.6788	0.7907	0.7569	0.7111	0.8117	0.7804	0.7377
2	100	0.8109	0.7821	0.7428	0.8275	0.8002	0.7628	0.8412	0.8156	0.7803
3	10	0.0743	0.0395	0.0160	0.4000	0.3117	0.2190	0.5800	0.4959	0.3973
3	15	0.1967	0.1389	0.0847	0.4454	0.3684	0.2814	0.5860	0.5141	0.4270
3	20	0.2972	0.2338	0.1661	0.4892	0.4197	0.3376	0.6028	0.5386	0.4591
3	25	0.3758	0.3128	0.2410	0.5278	0.4641	0.3867	0.6217	0.5632	0.4896
3	30	0.4379	0.3776	0.3057	0.5612	0.5024	0.4292	0.6402	0.5863	0.5175
3	40	0.5297	0.4758	0.4084	0.6153	0.5642	0.4988	0.6737	0.6268	0.5657
3	50	0.5942	0.5463	0.4846	0.6570	0.6119	0.5530	0.7021	0.6604	0.6053
3	75	0.6948	0.6579	0.6087	0.7286	0.6936	0.6467	0.7551	0.7223	0.6782
3	100	0.7534	0.7236	0.6832	0.7741	0.7456	0.7067	0.7915	0.7645	0.7275
4	10	0.0299	0.0132	0.0041	0.3360	0.2515	0.1673	0.5357	0.4478	0.3485
4	15	0.1239	0.0819	0.0455	0.3753	0.3022	0.2225	0.5313	0.4583	0.3721
4	20	0.2154	0.1631	0.1099	0.4182	0.3516	0.2752	0.5449	0.4803	0.4019
4	25	0.2923	0.2370	0.1760	0.4575	0.3959	0.3227	0.5631	0.5043	0.4316
4	30	0.3558	0.3006	0.2368	0.4924	0.4350	0.3650	0.5819	0.5277	0.4595
4	40	0.4530	0.4015	0.3383	0.5504	0.4998	0.4360	0.6174	0.5700	0.5090
4	50	0.5235	0.4764	0.4169	0.5962	0.5509	0.4926	0.6483	0.6059	0.5506
4	75	0.6366	0.5992	0.5497	0.6765	0.6408	0.5933	0.7076	0.6739	0.6289
4	100	0.7041	0.6733	0.6319	0.7287	0.6992	0.6593	0.7492	0.7212	0.6832

Table 11.3 *(continued)*

k	n/α	d.f. 0			d.f. 10			d.f. 20		
		0.05	0.01	0.001	0.05	0.01	0.001	0.05	0.01	0.001
5	10	0.0096	0.0033	0.0007	0.2906	0.2083	0.1308	0.5072	0.4139	0.3123
5	15	0.0763	0.0470	0.0236	0.3214	0.2518	0.1786	0.4891	0.4146	0.3290
5	20	0.1560	0.1138	0.0727	0.3618	0.2982	0.2271	0.4984	0.4333	0.3559
5	25	0.2283	0.1806	0.1296	0.4006	0.3413	0.2723	0.5151	0.4561	0.3843
5	30	0.2906	0.2412	0.1852	0.4359	0.3803	0.3136	0.5336	0.4792	0.4117
5	40	0.3895	0.3412	0.2828	0.4960	0.4463	0.3844	0.5699	0.5222	0.4615
5	50	0.4634	0.4182	0.3616	0.5443	0.4994	0.4421	0.6024	0.5596	0.5042
5	75	0.5854	0.5482	0.4994	0.6310	0.5950	0.5473	0.6662	0.6319	0.5863
5	100	0.6600	0.6289	0.5872	0.6885	0.6584	0.6180	0.7120	0.6832	0.6444
6	15	0.0451	0.0254	0.0112	0.2790	0.2121	0.1445	0.4562	0.3798	0.2943
6	20	0.1123	0.0784	0.0472	0.3160	0.2550	0.1887	0.4602	0.3945	0.3181
6	25	0.1786	0.1375	0.0949	0.3535	0.2964	0.2313	0.4748	0.4156	0.3448
6	30	0.2382	0.1938	0.1448	0.3886	0.3346	0.2710	0.4925	0.4381	0.3715
6	40	0.3364	0.2910	0.2371	0.4495	0.4007	0.3407	0.5289	0.4811	0.4209
6	50	0.4121	0.3685	0.3146	0.4995	0.4550	0.3988	0.5624	0.5193	0.4641
6	75	0.5403	0.5033	0.4551	0.5909	0.5546	0.5070	0.6296	0.5948	0.5488
6	100	0.6205	0.5890	0.5471	0.6526	0.6220	0.5811	0.6787	0.6493	0.6099
7	15	0.0249	0.0127	0.0049	0.2463	0.1821	0.1194	0.4331	0.3543	0.2685
7	20	0.0794	0.0535	0.0304	0.2786	0.2207	0.1592	0.4295	0.3636	0.2882
7	25	0.1389	0.1045	0.0700	0.3141	0.2597	0.1989	0.4410	0.3822	0.3128
7	30	0.1949	0.1562	0.1143	0.3481	0.2966	0.2367	0.4573	0.4033	0.3381
7	40	0.2907	0.2493	0.2007	0.4088	0.3617	0.3044	0.4929	0.4455	0.3864
7	50	0.3668	0.3261	0.2761	0.4596	0.4163	0.3620	0.5266	0.4839	0.4294
7	75	0.4992	0.4635	0.4172	0.5542	0.5184	0.4716	0.5960	0.5612	0.5155
7	100	0.5838	0.5530	0.5120	0.6192	0.5888	0.5481	0.6477	0.6182	0.5786
8	20	0.0546	0.0347	0.0182	0.2474	0.1913	0.1335	0.4047	0.3373	0.2618
8	25	0.1070	0.0780	0.0498	0.2805	0.2279	0.1703	0.4125	0.3530	0.2841
8	30	0.1591	0.1247	0.0884	0.3133	0.2632	0.2060	0.4270	0.3727	0.3081

Table 11.3   *(continued)*

d.f.		0			10			20		
k	n/α	0.05	0.01	0.001	0.05	0.01	0.001	0.05	0.01	0.001
8	40	0.2517	0.2130	0.1682	0.3732	0.3270	0.2715	0.4611	0.4136	0.3550
8	50	0.3275	0.2884	0.2409	0.4243	0.3815	0.3284	0.4947	0.4518	0.3977
8	75	0.4628	0.4273	0.3817	0.5212	0.4853	0.4387	0.5656	0.5305	0.4845
8	100	0.5510	0.5199	0.4789	0.5889	0.5581	0.5172	0.6193	0.6193	0.5494
9	20	0.0365	0.0217	0.0104	0.2210	0.1666	0.1123	0.3838	0.3146	0.2391
9	25	0.0818	0.0576	0.0351	0.2513	0.2005	0.1462	0.3874	0.3274	0.2591
9	30	0.1294	0.0992	0.0682	0.2827	0.2342	0.1798	0.3999	0.3455	0.2816
9	40	0.2178	0.1820	0.1413	0.3414	0.2963	0.2428	0.4324	0.3850	0.3270
9	50	0.2924	0.2553	0.2107	0.3924	0.3503	0.2986	0.4658	0.4228	0.3691
9	75	0.4290	0.3943	0.3499	0.4909	0.4551	0.4089	0.5377	0.5023	0.4564
9	100	0.5200	0.4891	0.4485	0.5608	0.5298	0.4888	0.5931	0.5627	0.5225
10	20	0.0230	0.0126	0.0054	0.1982	0.1455	0.0947	0.3671	0.2961	0.2204
10	25	0.0611	0.0414	0.0239	0.2258	0.1769	0.1258	0.3658	0.3055	0.2379
10	30	0.1043	0.0780	0.0518	0.2557	0.2089	0.1574	0.3761	0.3218	0.2589
10	40	0.1881	0.1551	0.1181	0.3130	0.2692	0.2178	0.4068	0.3596	0.3026
10	50	0.2612	0.2260	0.1842	0.3636	0.3225	0.2723	0.4396	0.3970	0.3439
10	75	0.3986	0.3645	0.3212	0.4632	0.4277	0.3822	0.5121	0.4768	0.4311
10	100	0.4920	0.4612	0.4209	0.5349	0.5039	0.4631	0.5688	0.5384	0.4982

This table is taken from Grubbs 1979 pages 3-69-71 by permission. Table adapted from Hawkins 1978. Chapter 3, Army Statistics Manual, DARCOM p. 706-1-3.

SKEWNESS AND KURTOSIS TESTS

These types of tests are more routinely used to test for the normality of data distribution. They are mentioned here, in the context of outlier determination, merely to point out that such tests are available.

The skewness test is used when suspect outliers are only on one side. The kurtosis test is generally recommended when suspect outliers are on both sides or when scalar contamination is suspected. The tests can be applied sequentially, until no more values are rejected. The articles by Ferguson (1961), as well as the critical value tables of Pearson (1965) or Mulholland (1977), should be consulted, for more information on this approach.

SLIPPAGE TESTS

The foregoing tests have been based upon the assumption that we are dealing with a normally distributed population. There are also available, however, distribution free or nonparametric tests that can be applied when the population is deemed to be not normally distributed (Neave, 1979). The advantages of such tests are their protection against non-normally distributed populations. However, for a normal population, these tests are not statistically as powerful as are the parametric tests.

k-Sample Slippage Test (Conover, 1971)

Let us consider the case where one has data from several "k" random samples. These samples are mutually independent of each other, as in the case of data generated in different laboratories conducting the same tests or analytical procedures.

The underlying assumptions are that:

1. Each sample is a random sample from some population;
2. The k samples are mutually independent;
3. The test will be exact for continuous random variables;
4. Either the k population distribution functions are identical, or some populations will tend to yield larger observations than other populations (a one-sided test); and
5. The data are ordinal in scale.

The test operates by comparing the "largest" sample with the "smallest" sample in the following manner. Determine the largest measurement in each of the k samples and mark it with an asterisk (*). The smallest of these extreme values is designated z^1 and the largest is designated z. The k-sample slippage test statistic Z equals the number of measurements from the sample containing z larger than the measurement z^1.

In the case where several of the k samples have the same z value, the second largest value should be used to break the tie. If necessary, one could even go to the third level, to designate the largest sample set z.

An example application follows:

1. Ten measurements selected at random from five separate (independent) laboratories (workplaces) are tabulated in Table 11.4.
2. Place an asterisk (*) beside the largest measurement, in each sample (workplace location).
3. Underline the *smallest largest* measurement once (177 from location E).
4. Under line the *largest largest* measurement twice (258 from location B).
5. Count the number of measurements from location B that exceed the *smallest largest* value (177). This is seven.
   Therefore, Z = 7.
6. From Table 11.5, the critical value for p = 0.95 is 5. Therefore, since the critical value is smaller than the calculated value, one concludes that laboratory (workplace) B tends to have higher measurements than laboratory E.
7. The test can also be applied sequentially. In this case, laboratory A would be tested with a k of 4. It would also be determined to have slipped. Testing laboratory D, with k = 3, yields the decision that it too has slipped. Laboratory C, with k = 3, does not appear to have slipped. This Conover k sample slippage test has been studied for its power. It was thought to be more powerful than the parametric F test, where the populations are normal, and these populations with larger means also have larger variances.

YOUDEN EXTREME RANK SUM TEST FOR OUTLIERS

A nonparametric extreme rank sum test was originally published in 1963. The mathematical development of this test was published by Thompson and Wilke, also in 1963. The cases of data generated by multiple measuring instruments or people

Table 11.4   Workplace Locations

Measurements	A	B	C	D	E
1	143	210	159	113	98
2	152	183	69	128	106
3	216	258*	64	240*	66
4	236	204	85	134	177*
5	120	252	126	205	93
6	191	132	148	234	68
7	244*	171	98	188	81
8	104	231	90	149	143
9	208	226	202*	223	160
10	179	104	182	217	128

Table 11.5.   Quantiles of the k-Sample Slippage Test Statistics[a]

	k = 2					k = 3					k = 4				
	p = .80	.90	.95	.98	.99	p = .80	.90	.95	.98	.99	p = .80	.90	.95	.98	.99
n=3	2	2				2	3								
4	2	3	3			3	3	4	4			3			
5	2	3	3	4		3	4	4	5	5	3	4	4	5	5
6	2	3	4	4	4	3	4	4	5	5	3	4	5	5	6
7	2	3	4	5	5	3	4	4	5	5	3	4	5	5	6
8	3	3	4	5	5	3	4	5	5	6	3	4	5	6	6
9	3	3	4	5	5	3	4	5	5	6	3	4	5	6	6
10	3	3	4	5	6	3	4	5	6	6	4	4	5	6	6
12	3	3	4	5	6	3	4	5	6	6	4	4	5	6	7
14	3	3	4	5	6	3	4	5	6	7	4	5	5	6	7
16	3	4	4	5	6	3	4	5	6	7	4	5	5	6	7
18	3	4	4	5	6	3	4	5	6	7	4	5	6	7	7
20	3	4	4	6	6	3	4	5	6	7	4	5	6	7	7
25	3	4	5	6	7	3	4	5	7	7	4	5	6	7	7
30	3	4	5	6	7	3	4	5	7	7	4	5	6	7	8
35	3	4	5	6	7	3	5	5	7	7	4	5	6	7	8
40	3	4	5	6	7	3	5	5	7	8	4	5	6	7	8
Approximation for n>40	3	4	5	6	7	4	5	6	7	8	5	5	6	8	9

[a] The entries in this table are p quantiles w_p, for selected values of p, of the slippage test statistic as defined in Section V. To obtain the p quantile, enter the portion of the table for the correct number of samples k, read across the row corresponding to the size n of each sample, to the entry w_p in the column headed with the desired value of p = 1 - α. Reject H_0 at the level x if the test statistic exceeds w_p.

taking readings at several workplace areas would be typical applications. Another application would involve evaluating the results of multiple readers on several asbestos filters, to develop and qualify the personnel to be used for the procedure.

The Youden example, adapted to an asbestos microscopic count, is shown below, in Table 11.6. The data presented would seem to indicate that Technician 2 is producing results that are lower than the other technicians.

This test is applicable to the case cited above and, of course, to the original work of Youden, where temperature reference cells (objects) were tested (judged) by several thermometers, or where laboratories (objects) were tested by several samples (judges).

The application of the test to the set of data in Table 11.6 follows:

1. Arrange the data, as in Table 11.6.
2. Rank the values for each "judge" (thermometer, filter, material being tested, etc.), in ascending order. Ties are broken by flipping a coin.
3. Sum up the ranks for each "object" being tested (chemical reference cell, technician, laboratory, etc.)
4. Compare the rank sums with the appropriate entry from Table 11.7. In this case, the approximate two-tail limits at 5% are 4 and 16, when considering results of four objects and four laboratories.
5. With this small sample, we reject the work of Technician 2, whose result (rank sum = 4) appears at the low limit of the two-tail test. The other technicians' results fall within the table limits.

This test is a distribution-free test, as is the Conover k-sample slippage test. Both these should find considerable application, where the laboratory results are not normally distributed.

SUMMARY

This chapter discussed several techniques for dealing with two or more outliers in a sample. The handling of several sets of data was covered, using the k-sample slippage test of Conover and Youden's extreme rank sums test.

Table 11.6   Asbestos Measurement Values

Technician	Filter				Rank Sum
	A	B	C	D	
1	36(4)	38(2)	36(2)	30(2)	10
2	17(1)	18(1)	26(1)	17(1)	4
3	30(3)*	39(3)	41(4)	34(4)	14
4	30(2)*	45(4)	38(3)	33(3)	12

*The tie was broken by flipping a coin.

Table 11.7. Approximate Five Percent Two Tail Limits for Ranking Scores

No. of Objects / No. of Labs.	Number of Judges / Number of Materials												
	3	4	5	6	7	8	9	10	11	12	13	14	15
3		4	5	7	8	10	12	13	15	17	19	20	22
		12	15	17	20	22	24	27	29	31	33	36	38
4		4	6	8	10	12	14	16	18	20	22	24	26
		16	19	22	25	28	31	34	37	40	43	46	49
5		5	7	9	11	13	16	18	21	23	26	28	31
		19	23	27	31	35	38	42	45	49	52	56	59
6	3	5	7	10	12	15	18	21	23	26	29	32	35
	18	23	28	32	37	41	45	49	54	58	62	66	70
7	3	5	8	11	14	17	20	23	26	29	32	36	39
	21	27	32	37	42	47	52	57	62	67	72	76	81
8	3	6	9	12	15	18	22	25	29	32	36	39	43
	24	30	36	42	48	54	59	65	70	76	81	87	92
9	3	6	9	13	16	20	24	27	31	35	39	43	47
	27	34	41	47	54	60	66	73	79	85	91	97	103
10	4	7	10	14	17	21	26	30	34	38	43	47	51
	29	37	45	52	60	67	73	80	87	94	100	107	114
11	4	7	11	15	19	23	27	32	36	41	46	51	55
	32	41	49	57	65	73	81	88	96	103	110	117	125
12	4	7	11	15	20	24	29	34	39	44	49	54	59
	35	45	54	63	71	80	88	96	104	112	120	128	136
13	4	8	12	16	21	26	31	36	42	47	52	58	63
	38	48	58	68	77	86	95	104	112	121	130	138	147
14	4	8	12	17	22	27	33	38	44	50	56	61	67
	41	52	63	73	83	93	102	112	121	130	139	149	158
15	4	8	13	18	23	29	35	41	47	53	59	65	71
	44	56	67	78	89	99	109	119	129	139	149	159	169

This table is taken from the book, *Statistical Manual of the Association of Official Analytical Chemists* by W. J. Youden and E. H. Steiner 1975. By permission.

References

Chauvenet, W. 1960. *A Manual of Spherical and Practical Astronomy: Vol II*. New York: Dover Publications (Unabridged and unaltered republication of the 5th revised and corrected edition [copyright 1891]).

Ferguson, T. S. 1961. On the rejection of outliers. In *Fourth Berkeley Symposium on Mathematical Statistics and Probability*. Ed. Jerzy Neyman, pp. 253-87. Berkeley and Los Angeles, CA: University of California Press.

Ferguson, T. S. 1961. Rules for the rejection of outliers. *Revue Institute Int. de Stat* Vol. 3:29-43

Grubbs, F. E. 1950. Sample criteria for testing outlying observations. *Annals of Mathematical Statistics* 21:27-58.

Grubbs, F. E. 1969. Procedures for detecting outlying observations in samples. *Technometrics*. 11:1-21.

Grubbs, F. E. 1979. Procedures for detecting outlying observations. In *Army Statistics Manual DARCOM-P706-1-3*. Chapter 3. Aberdeen Proving Grounds, MD: U.S. Army Aberdeen Research and Development Center.

Grubbs, F. E. and G. Beck. 1972. Extension of sample sizes and percentage points for significance tests of outlying observations. *Technometrics*. Vol. 14, No. 4:847-854.

Hawkins, D. M. 1978. Fractiles of an extended multiple outlier test. *Journal of Statistical Computation and Simulation*. Manuscript submitted for publication.

McMillan, R. G. 1971. Tests for one or two outliers in normal samples with unknown variance. *Technometrics*. Vol. 13, No. 1:87-100

McMillan, R. G. and H. A. David. 1971. Tests for one or two outliers in normal samples with known variance. *Technometrics*. Vol. 13, No. 1:75-85.

Moran, M. A. and R. G. McMillan. 1973. Tests for one or two outliers in normal samples with unknown variance: a correction. *Technometrics*. Vol. 15, No. 3:637-40.

Mulholland, M. P. 1977. On the null distribution of $\sqrt{b_1}$ for samples of size at most 25, with tables. *Biometrika*. Vol. 52:282-85.

Neave, H. R. 1979. Quick and simple tests based on extreme observations. *Journal of Quality Technology*. Vol. 11, No. 2:66-79.

Pearson, E. S. Tables of percentage points of $\sqrt{b}$ and b_2 in normal samples; a rounding off. *Biometrika*. Vol. 52:282-85.

Rosner, B. 1975. On the detection of many outliers. *Technometrics*. Vol. 17, No. 2:221-27.

Rosner, B. 1977. Percentage points for the RST many outlier procedure. *Technometrics*. Vol. 19, No. 3:307-12.

Thompson, W. A. and T. A. Willke. 1963. On an extreme rank sum test for outliers. *Biometrika*. 50:375-83.

Tietzen, G. L. and R. H. Moore. 1972. Some Grubbs-type statistics for the detection of several outliers. *Technometrics*. Vol. 14, No. 3:583-97.

Youden, W. J. 1963: Ranking laboratories by round robin tests. *Material Research and Standards*. Vol. 3:9-13.

Youden, W. J., and E. H. Steiner. 1975. *Statistical Manual of the Association of Official Analytical Chemists*. Washington, DC: Association of Official Analytical Chemists.

12

Outliers Appearing in Continuous Data Sets

INTRODUCTION

The preceding chapters have covered outlier rejection criteria and techniques, for a single set of data, or the rejection of a single set of data, from several sets of data. The techniques of Conover (k-sample slippage test) and Youden (extreme rank sum test) are examples of the latter. There are cases and processes, however, that produce data either continually or continuously. It is these latter situations that will be discussed in this chapter.

DUPLICATE MEASUREMENTS

Replicate determination rejection rules were discussed by Anscomb in 1960. Tietjen and Beckman (1974) presented a comprehensive discussion of the analytical laboratory problem of dealing with duplicate measurements. If the two results do not agree, what action should be taken? If a third analysis is run, how is its result handled? The same problem arises when two laboratories perform a test or analysis. If the results disagree, and a third laboratory runs a similar test or analysis, what is to be done with the result of the referee laboratory?

Various procedures that may be used in case of disagreement between two original results are outlined below:

A. Use only the third analysis (or referee analysis or test) value, as the reported or accepted value.
B. Use the third analysis result (or referee analysis), if it falls within the first two results in question. If not, accept the value closest to the referee value. This procedure simply accepts the median value.

C. Use the average of the third value and the closest of the two original values.

D. Use the average of the closest pair of the three values.

E. Use the average of all three observations.

F. Use the average of the lowest pair of values. (This procedure is used when outliers, because of contamination, can occur only on the high side.)

G. Use the average of the lowest and highest values.

The decision to do a third analysis or test is in itself a rejection of outlier decision. Tietjen and Beckman recommend a Grubbs single sample test, with some prior knowledge of the variability.

Tietjen and Beckman conclude that:

1. Procedures E and G are the worst choice because of heavy bias.
2. Procedure F is best, if one can expect outliers to occur only on one side.
3. Procedures C and D are the best choice and are practically indistinguishable.
4. Procedures A and B are intermediate choices between the best (C and D) and the worst (E and G).

They conclude their studies with the statement, "The final conclusion, restated because it is not in accord with common practice, is, that with no prior knowledge of where the outliers will occur, and, given that the first two observations are not in agreement, it is uniformly better to report the mean of the closest pair than either the median or the referee's value."

ROUTINE TRIPLICATE MEASUREMENTS

Tietjen and Beckman summarize the arguments of Youden (1949) and Lieblein (1952). The three measurements are a random sample from a normal population. The common practice of taking the mean of the "closest two" is condemned. The approved technique is to test for an outlier. If one is found at, say, the $\alpha \leq 0.01$ level, average the two remaining observations.

CONTROL CHARTS

The Shewhart control chart is a classic example of outlier technique. The $\overline{X}$ chart is specifically designed to detect values arising from a location change, in the basic population being monitored. The range or R chart is specifically designed to detect a change in population variability or spread, through the increased range of the sample values.

Since the "out-of-control" points of the control chart are, by definition, "outliers," no more needs to be said about this basic technique covered in earlier chapters. Standard texts such as Grant (1964) and the ASTM monograph on control charts can be consulted.

USE OF RUNS

A run is defined as a succession of values of the same type, where the values are given a two-way classification, such as above-the-average or below-the-average, or increasing or decreasing.

The following set of values (adopted from Duncan [1959]) will serve as an example:

Table 12.1

Order of Sample	Value	Order of Sample	Value	Order of Sample	Value
In time		In time		In time	
1	13	13	8	25	10
2	10	14	2	26	11
3	14	15	5	27	16
4	9	16	8	28	11
5	6	17	9	29	9
6	9	18	10	30	4
7	16	19	17	31	9
8	12	20	12	32	20
9	5	21	14	33	18
10	9	22	27		
11	22	23	17		
12	12	24	8		

These values are plotted below on Figure 12.1.

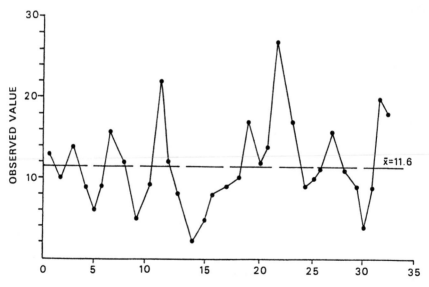

FIGURE 12.1

The types of runs that appear in this set of data include "runs above average," "runs below average," "run up," and "run down." Since the "run up" is a succession of increasing values, the data points 16, 17, 18, and 19 appear to be a "run up." Points 12, 13, and 14 are a "run down." Data points 19, 20, 21, 22, and 23 are a "run above average," whereas points 13, 14, 15, 16, 17, and 18 are a "run below average."

Two approaches are taken in studying runs. The total number of runs of a given type or the longest run of a type are both useful. The total number of runs of a given type, from a random series, have been studied by Swed and Eisenhart (1943), Stevens (1939), and Wald and Wolfowitz (1940). The test is run as follows (Duncan [1959]):

1. After examining the plot of Figure 12.1, the question to be answered is: Is the presence of two long runs below the mean line indicative of a biased situation (one that may contain outliers)?

2. Count the runs of one type, (e.g., runs above the mean line):

Run Type	Positions	# of Runs
Runs of 1	(1), (3), (27)	3
Runs of 2	(7, 8), (11, 12), (32, 33)	3
Runs of 5	(19, 20, 21, 22, 23)	$\frac{1}{7}$

3. Count the runs of the other type (runs below the mean line):

Run Type	Positions	# of Runs
Runs of 1	(2)	1
Runs of 2	(9, 10)	2
Runs of 3	(4, 5, 6), (24, 25, 26)	2
Runs of 4	(28, 29, 30, 31)	1
Runs of 6	(13, 14, 15, 16, 17, 18)	$\frac{1}{7}$

4. Let the smaller number of values (data points) of one type equal m. In this case, the number of data points above the line is 14, so m = 14.

5. Let the larger number of data points (of the other type) equal n. In this example, the number of data points below the mean line is 19, so n = 19.

6. Determine the total number of runs of each type. In this case:
   Runs above the mean line = 7
   Runs below the mean line = 7
   Total number of runs observed = 14

7. The values of m and n are used to enter Tables 12.2 or 12.3, at either the P = 0.005 or P = 0.05 level, to determine the critical values. If the observed total number of runs is less than the critical value, we can conclude that the number of runs is less that one would expect, based on the assumption of randomness. In this example, the observed number of runs (14) is more than the table

Table 12.2.   **Table for Testing Randomness of Grouping in a Sequence of Alternatives**
(Probability of an equal or smaller number of runs than that listed is P = 0.005). n equals cases on one side of average—m equals cases on other side of average. m is always the smaller number of cases, n the larger.

n\ ^{m}	6	7	8	9	10	11	12	13	14	15	16	17	18	19	20
6	2														
7	2	3													
8	3	3	3												
9	3	3	3	4											
10	3	3	4	4	5										
11	3	4	4	5	5	5									
12	3	4	4	5	5	6	6								
13	3	4	5	5	5	6	6	7							
14	4	4	5	5	6	6	7	7	7						
15	4	4	5	6	6	7	7	7	8	8					
16	4	5	5	6	6	7	7	8	8	9	9				
17	4	5	5	6	7	7	8	8	8	9	9	10			
18	4	5	6	6	7	7	8	8	9	9	10	10	11		
19	4	5	6	6	7	8	8	9	9	10	10	10	11	11	
20	4	5	6	7	7	8	8	9	9	10	10	10	11	12	12

critical values (9 at P = 0.005 or 12 at P = 0.05). Therefore, we conclude that the data set is randomly distributed by chance alone.

The same approach to examining the runs up and runs down finds that the number of increasing values equals 17 and that the number of decreasing values equals 15. The number of runs up is 7, and the number of runs down is 8. The total number of runs is, therefore, 15. Entering Tables 12.2 and 12.3 for m = 15 and n = 17, we find the critical values to be 9 at P = 0.005 and 11 at P = 0.05. Again, we conclude that the data set is randomly distributed.

An extension of Tables 12.2 and 12.3, for larger values of m and n, is provided in Table 12.4.

The second approach to studying runs is the "length of run" of a specific type. The technique involves using the median of the set of data. In case there are multiple values of the median, these points are assigned to the side with the smaller number of values, to balance the values on both sides. For an odd number of median values, one is ignored.

Having divided the data as above, determine the longest run on either side of the median, and determine whether it is longer than the critical values of Table 12.6. If the critical values of Table 12.6 are exceeded, we conclude that nonrandom factors are affecting this population.

In the case of the earlier set of data, the analysis would proceed as follows:

Table 12.3.   **Table for Testing Randomness of Grouping in a Sequence of Alternatives (Probability of an equal or smaller number of runs than that listed is P = 0.05). n equals cases on one side of average—m equals cases on other side of average. m is always the smaller number of cases, n the larger.**

$n\backslash^{m}$	6	7	8	9	10	11	12	13	14	15	16	17	18	19	20
6	3														
7	4	4													
8	4	4	5												
9	4	5	5	6											
10	5	5	6	6	6										
11	5	5	6	6	7	7									
12	5	6	6	7	7	8	8								
13	5	6	6	7	8	8	9	9							
14	5	6	7	7	8	8	9	9	10						
15	6	6	7	8	8	9	9	10	10	11					
16	6	6	7	8	8	9	10	10	11	11	11				
17	6	7	7	8	9	9	10	10	11	11	12	12			
18	6	7	8	8	9	10	10	11	11	12	12	13	13		
19	6	7	8	8	9	10	10	11	12	12	13	13	14	14	
20	6	7	8	9	9	10	11	11	12	12	13	13	14	14	15

Table taken from Swed, F. S. & Eisenhart, C. "Tables for Testing Randomness of Grouping in a Sequenct of Alternatives." Annals of Mathematical Statistics, Vol. XIV, 1943. By permission of The Institute of Mathematical Statistics.

1. Review the data from Table 12.1, and determine the median and assign multiple values to the side having fewer values. To do this, set up Table 12.5.

   The median value is 10, as there are 14 positions with values of 9 or less, and 16 with values of 11 or more. The values of positions 2 and 25 were assigned to the low side, as the runs are short. The 10 values at position 18 is left unassigned.

2. Determine the longest run on either side. The values for positions (13, 14, 15, 16, 17, and 18) and (19, 20, 21, 22, and 23) are 6 and 5 values long respectively.
3. Consult Table 12.6, to determine that the observed run lengths exceed the critical values. The total number of points being considered (30) is the n used in Table 12.6.
4. Since the critical value for n = 30, at the P = 0.05 level, is 8, no evidence of nonrandomness is present.

   The final test for nonrandomness by runs involves studying runs up or down. The approach is similar to the preceding case.

1. Determine the length of the longest run up or down. The values for positions (15, 16, 17, 18, and 19) represent the longest run (5).
2. Consult Table 12.7, to determine the probability of a given run length for the total number of values in the data set.

Table 12.4   Limiting Values for the Total Number of Runs Above and Below the Median of a Set of Values

Probability of an Equal or Smaller Value			Probability of an Equal or Smaller Value			Probability of an Equal or Smaller Value		
$m=n$	0.005	0.05	$m=n$	0.005	0.05	$m=n$	0.005	0.05
10	4	6	40	29	33	70	55	60
11	5	7	41	29	34	71	56	61
12	6	8	42	30	35	72	57	62
13	7	9	43	31	35	73	57	63
14	7	10	44	32	36	74	58	64
15	8	11	45	33	37	75	59	65
16	9	11	46	34	38	76	60	66
17	10	12	47	35	39	77	61	67
18	10	13	48	35	40	78	62	68
19	11	14	49	36	41	79	63	69
20	12	15	50	37	42	80	64	70
21	13	16	51	38	43	81	65	71
22	14	17	52	39	44	82	66	71
23	14	17	53	40	45	83	66	72
24	15	18	54	41	45	84	67	73
25	16	19	55	42	46	85	68	74
26	17	20	56	42	47	86	69	75
27	18	21	57	43	48	87	70	76
28	18	22	58	44	49	88	71	77
29	19	23	59	45	50	89	72	78
30	20	24	60	46	51	90	73	79
31	21	25	61	47	52	91	74	80
32	22	25	62	48	53	92	75	81
33	23	26	63	49	54	93	75	82
34	23	27	64	49	55	94	76	83
35	24	28	65	50	56	95	77	84
36	25	29	66	51	57	96	78	85
37	26	30	67	52	58	97	79	86
38	27	31	68	53	58	98	80	87
39	28	32	69	54	59	99	81	87
						100	82	88

Swed & Eisenhart.
Used by permission of the Institute of Mathematical Statistics.

3. Since the probability of a run of 5 for n = 33 appears only to approach the P = 0.05 level, one cannot conclude that the sample is nonrandom.

CONCLUSION

Rules have been provided, for properly considering duplicate and triplicate readings from instrument or laboratory results. How to properly handle values after rejection have been covered as well. Control charts have been shown to be a natural outlier detection and rejection technique, specifically designed for the purpose in continuous data generating processes. Using runs to identify nonrandomness provides an effective nonparametric technique.

Table 12.5

Sample Order	Value	H or L Side
1	13	H
2	10	M-L
3	14	H
4	9	L
5	6	L
6	9	L
7	16	H
8	12	H
9	5	L
10	9	L
11	22	H
12	12	H
13	8	L
14	2	L
15	5	L
16	8	L
17	9	L
18	10	M-
19	17	H
20	12	H
21	14	H
22	27	H
23	17	H
24	9	L
25	10	M-L
26	11	H
27	16	H
28	11	H
29	9	L
30	4	L
31	9	L
32	20	H
33	18	H

**Table 12.6   Critical Values for Lengths of Runs on Either Side of the Median of N Cases.
(Probability of getting at least one run of specified size or more)**

N	0.05	0.01	0.001
10	5	—	—
20	7	8	9
30	8	9	—
40	9	10	12
50	10	11	—

F. Mosteller, "Note on Application of Runs to Quality Control Charts." Annals of Mathematical Statistics, Vol. XII, 1941, p. 232. By permission. of the Institute of Mathematical Statistics.

Table 12.7   Limiting Values for Lengths of Runs Up and Down in a Series of N Numbers

	Probability Equal to or Less Than 0.0032		Probability Equal to or Less Than 0.0567	
N	Run	Probability of an Equal or Greater Run	Run	Probability of an Equal or Greater Run
4	4	0.0028	4	0.0028
5	5	0.0004	4	0.0165
6	5	0.0028	4	0.0301
7	6	0.0004	4	0.0435
8	6	0.0007	4	0.0567
9	6	0.0011	5	0.0099
10	6	0.0014	5	0.0122
11	6	0.0018	5	0.0146
12	6	0.0021	5	0.0169
13	6	0.0025	5	0.0193
14	6	0.0028	5	0.0216
15	6	0.0032	5	0.0239
20	7	0.0006	5	0.0355
40	7	0.0015	6	0.0118
60	7	0.0023	6	0.0186
80	7	0.0032	6	0.0254
100	8	0.0005	6	0.0322
200	8	0.0010	7	0.0085
500	8	0.0024	7	0.0215
1,000	9	0.0005	7	0.0428
5,000	9	0.0025	8	0.0245

This table is taken from P. S. Olmstead, "Distribution of Sample Arrangements for Runs Up and Down." *Annals of Mathematical Statistics*, Vol. XVII (1946) p. 29. By permission. of the Institute of Mathematical Statistics

REFERENCES

Anscombe, F. J. 1960. Rejection of outliers. *Technometrics* 2(2):123–47.

———. 1951. *Manual on Quality Control of Materials*. Philadelphia, PA: American Society for Testing Materials.

Duncan, Acheson J. 1959. *Quality Control and Industrial Statistics*. Homewood, IL: Richard D. Irwin.

Grant, E. L. and R. S. Leavenworth. 1964. *Statistical Quality Control*. New York: McGraw-Hill Book Company.

Lieblein, J. 1952. Properties of certain statistics involving the closest pair in a sample of three observations. *Journal of Research National Bureau of Standards*. 48:255–68.

Stevens, W. L. 1939. Distribution of groups in a sequence of alternatives. *Annals of Eugenics* IX(I)10–17.

Swed, S., and C. Eisenhart. 1943. Tables for testing randomness of grouping in a sequence of alternatives. *Annals of Mathematical Statistics* XIV:147–62.

Tietjen, G. L., and R. J. Beckman. 1974. Duplicate measurements in the chemical laboratory. *Technometrics* 16(4)53–56.

Wald, A., and J. Wolfowitz. 1940. On a test whether two samples are from the same population. *Annals of Mathematical Statistics*. XI:147–62.

13

Statistical Quality Control for Asbestos Counting

INTRODUCTION

Supplemental to the basic statistical techniques described in the foregoing chapters, it is perhaps useful to discuss how these methods are put into use, in a specific testing or analytical operation—asbestos analysis.

Laboratory quality control is a relatively new field, when compared with the quality control measures adapted by manufacturers over 50 years ago. It is also a dynamic one, undergoing rapid, continuous development. Generally, the selection and use of these procedures is left up to individual laboratory technicians, chemists, supervisors, or directors. This practice, in and of itself, is not necessarily bad, but it does lead to fragmentation and inconsistencies in the quality control of laboratory output.

Quality control requirements for laboratories performing analyses, as a result of some type of regulatory or accrediting body, are becoming more and more common. These requirements, generally rudimentary, now exist in the fields of occupational safety and health, medicine, and environmental sciences, as well as in physical testing work. However, the mere existence of a legal requirement does not necessarily relieve laboratory personnel of a decision-making burden. The laboratory staff may still need to select the exact statistical and analytical or testing procedures that will be used in complying with whatever requirement is at issue. Legal regulations may call for varying levels of quality control, thus making the quality process more difficult. For example, the legal requirement may be based on a simple internal quality control procedure (an intra-laboratory program), or it may call for a more complex inter-laboratory quality control program.

AVAILABLE QUALITY CONTROL PROCEDURES

Readily available statistical quality control procedures can be divided into two broad categories: those that are designed to maintain an acceptable level of variability (usually measured by the variance) in laboratory performance, and those that require the laboratory to produce acceptable results on standard reference samples. A functional quality control program may actually use elements from both types of procedures.

How to maintain an acceptable level of variability is typically a subjective matter. There is little published guidance given in this area. What there is results from someone's subjective opinion. A typical quality control program might incorporate some measure of within or intra-laboratory variability, such as the routine analysis of samples that have already been analyzed in that laboratory, and some measure of good performance on standard samples.

The goal of the variability measurement is to aid the laboratory, in evaluating its own performance over a period of time. The goal of any laboratory performing testing or chemical analysis should be to reduce the measured variability in the results.

Procedures in the laboratory. Some of the elements of good practice include the better training of personnel, clearer exposition of analytical or testing procedures, and correct and timely maintenance of equipment. These are all basic considerations, but they still need to be mentioned because so many facilities fail to implement even basic procedural practices, which improves the quality of analytical or test results.

The concept of the analytical laboratory is undergoing some reevaluation at this time. There are many newly established facilities that bear the name "laboratory," but they do not offer the full scope of services that the traditional testing or analytical laboratory is capable of offering.

Examples of these types of facilities are laboratories that do nothing but asbestos analyses or radon analyses. These laboratories are often operated by individuals who have little experience in analytical chemistry or laboratory management. They may be staffed by people who are trained only in the methods and procedures of the laboratory's specialty. Their understanding of the need for basic good laboratory procedures, including statistical quality control, is limited. Good laboratory practices include properly using blanks and reference samples (when applicable), cleanliness, training, rugged proven methods, availability of adequate reference works, and maintenance and calibration.

TRADITIONAL TECHNIQUES

Certain time-honored techniques have been used in laboratory quality control, for many years. Control charting, spiking, and using standard reference samples are all traditional quality control procedures. There is nothing wrong with these approaches; they are all useful within their own scope of application. However, there seems to have been a turn away from these traditional approaches, to more "exotic" approaches, during the last decade.

In some cases, laboratories are required to participate in a quality control program administered by a non-biased third party, such as a professional or technical society or association, or perhaps a government agency. Such programs are administered by The American Association for Laboratory Accreditation, the U.S. Environmental Protection Agency, the National Institute for Science and Technology, The American Industrial Hygiene Association, and many others. The scope and cost of these programs generally limit the number of materials involved in the program and the number of samples of each material to a minimum level. Because of this, there can be little doubt that, over a period of time, some laboratories have developed a "feel" for what answers are expected by the administering organization. This sort of shortcut saves the laboratory some money, but provides no benefits to the clients of the laboratory.

There is a tendency on the part of laboratories to misuse the evidence of successful participation in a proficiency testing program. In order to be considered an acceptable performer in one of these programs, a laboratory may be required to successfully analyze only a limited number of samples. Legally, the fact that a laboratory appears to be capable of analyzing an air filter sample for lead has no probative value with respect to its ability to analyze a soil sample for PCBs. Consequently, any laboratory doing work with analytes other than those for which it submits data as an element of a proficiency testing program must be able to provide adequate quality control information on all of its work. Few laboratories are capable of doing this today.

A typical example of a legislated quality control requirement is found in the U.S. Department of Labor regulation, Title 29, Part 1910.1001 of the Code of Federal Regulations. This standard, promulgated by the Occupational Safety and Health Administration, applies to all occupational exposures to asbestos, tremolite, anthrophyllite, and actinolite, except in construction work. It provides requirements for quality control procedures that should be employed by laboratories performing asbestos analyses, as stipulated by provisions of the asbestos standard. The regulation requires laboratories that perform asbestos analyses, in conjunction with asbestos work done to meet provisions of the standard, to implement a variety of quality control practices. Included among these are the routine analysis of acceptance blanks, laboratory blanks, and field blanks, comprehensive training of the analysts in specific analytical procedures, use of equipment that meets rigorous tolerance specifications, and the analysis of blind samples. In addition to these internal laboratory requirements, there is a requirement that every laboratory performing asbestos analyses, in compliance with the provisions of the standard, participate in an inter-laboratory quality control program that tests the proficiency of the laboratory in performing asbestos analysis.

The regulation, in 29CFR1910.1001, Appendix A, requires each laboratory that analyzes asbestos, tremolite, actinolite, and anthophyllite samples, for compliance determinations, to implement an inter-laboratory quality assurance program that, as a minimum, includes the participation of at least two other laboratories. Each laboratory is required to participate in round-robin testing, at least once every six months, with at least all the other laboratories in its quality assurance testing group. Each laboratory is required to submit slides (samples) that are typical of its workload for use in the program. The regulation specifies that the round-robin program be designed and the results analyzed, using appropriate statistical methodology. No further guidance is provided by the regulation; it is left to the discretion of the laboratory personnel to devise a program that will meet the specifications of the regulation.

This kind of approach to quality control might be characterized as "subjective quality control," since there are no objective asbestos standard samples. The true value of an asbestos sample is determined by a consensus analysis. There is only one method of analysis—optical microscopy. There are no "absolute" methods that can be used to completely characterize any sample parameter, in the way that neutron activation analysis might be used, when dealing with a metal sample. There are few methods available for handling the control of quality in this instance.

One method for dealing with this problem uses a statistical procedure for determining the pooled variance, as represented by the coefficient of variation (CV) (Abell et al.) This procedure attempts to evaluate the quality of fiber count data. The asbestos analytical procedure involves using an optical microscopic technique to count the number of fibers collected on a membrane filter. This procedure is highly subjective and depends on, among other things, the visual acuity, training, motivation, and experience of the analyst.

In this procedure, three critical parameters are calculated: the intra-counter CV, which is a measure of the ability of a single analyst to reproduce previous results; the intra-laboratory CV, which is a measure of the variability inherent in analytical results from the same sample read by more than one analyst; and the inter-laboratory CV, which is the measure of the ability of separate laboratories to produce similar analytical results from the same sample.

The calculations required to determine these values are complex, and the results are not clear-cut. There is no guidance given to the user of this procedure, as to what constitutes acceptable values for any of these parameters. Obviously, the lower the numerical value found, the better the method is deemed to be performing. In reality, the acceptability of the numbers is determined by what people are willing to accept. Asbestos analytical results are frequently used to determine conformance to a regulatory requirement. When this is done, the variability in the sample result must be taken into account. Typically, this means that the end user will calculate confidence limits around the result, to indicate the possible range for the true value of the sample. The confidence limit interval, or spread, is a function of the variability inherent in the analytical method—the greater the variability, the greater the confidence interval. If the user employs the upper and lower confidence limits for decision making, rather than using the sample value itself, the magnitude of the analytical variability can have a significant effect on the ability of the user to make a decision.

If the user can tolerate the variability produced by the analytical laboratory, it really makes no difference what that variability is. However, intuitively, small variabilities seem to be more desirable than larger ones. Therefore, one important benefit that derives from laboratory statistical quality control programs is a measure of the variability in the laboratory's results. Knowledge of this variability allows the users of the data to more accurately evaluate the situation that served as the source of the data.

The procedure developed by Abell et al., and mentioned earlier, is designed to enable the calculation of an inter-laboratory coefficient of variation. This inter-laboratory CV can then be used to calculate the upper and lower confidence limits around the sample value. Since the asbestos analysis is subjective, there are no absolute standard values to which the calculated intra-counter, intra-laboratory, or inter-laboratory CVs can be compared. Abell et al. mention that laboratories with good quality assurance programs have been able to achieve an intra-laboratory coefficient of variation between 0.17 and 0.22. The authors report that inter-laboratory coefficients of variation can be expected to be approximately twice the intra-laboratory coefficient of variation.

The procedure for calculating these CV values follows. See Table 13.1 for an example of the calculations.

1. First select 15 or more actual samples in each of three analytical count ranges: 5 to 20 fibers, 20.5 to 50 fibers, and more than 50 fibers. These represent the

Table 13.1   Intracounter CV Determination (High Range)

Original Count	Second	Third	$\overline{X}$	s	CV
338	386	371	365	24.5	0.067
719	637	411	589	159.5	0.270
535	335	328	349	117.5	0.294
227	276	197	233	39.8	0.170
318	428	366	370	55.1	0.149

Note 1: All fiber counts are in fibers per square millimeter.
Note 2: Only 5 samples were used to illustrate the calculation, even though 15 or more are recommended.
Note 3: The data used in this table are real data, produced by several counters at the same laboratory.

number of fibers counted by the analyst, in 100 microscope fields or less.

2. Calculate the fiber density for each sample, in terms of the fiber count per square millimeter on the surface of the filter.

3. Recount each of the samples, using the original counter (intra-counter) and counter(s) in the same laboratory (intra-laboratory) or another laboratory (inter-laboratory).

4. Compute the sample averages, $\overline{X}$; standard deviations, s, and coefficients of variation, $CV = s/\overline{X}$.

5. Compute the sum of the squares of the CV values.

6. Compute the pooled CV using the formula:

$$CV \text{ (pooled)} = \sqrt{\frac{CV_1^2 + CV_2^2 + .., + CV_n^2}{n}}$$

7. The evaluation of the magnitude of the pooled CV value is somewhat subjective. Abel et al. report that competent laboratories, working with optimally loaded samples, can achieve an intra-laboratory CV, somewhere between 0.17 and 0.22.

$$\Sigma CV^2 = 0.214$$

$$CV \text{ (pooled)} = \sqrt{\frac{0.214}{5}} = 0.21$$

The pooled CV for the high count range for the counters in this laboratory, using only five samples, is 0.21, which is an acceptable value for this type of analytical work.

In addition to calculating the coefficient of variation for fiber counts, another way to check the quality of microscopic counting of fibers is the sample quality test presented in the latest revision of NIOSH Method 7400. This sample quality test evaluates recounts by the same counter on a single sample. If the difference between two counts is very large, and the recounts fail the test, the failure is indicative of some problem with the counter's performance, or with the sample itself. In either case, if a pair of counts fails the test, the data is suspect and must be used judiciously, if it is used at all.

Implementing this test requires that the pooled intra-laboratory CV calculated above be converted to the square root scale. This is done simply by dividing the pooled CV by 2. Once this has been done, the following test is used: If

$$[y_1 - y_2] > (y) (CV)$$

where - CV = pooled square root scale value, then the counts fail the test,
and y_1 = First count on sample
      y_2 = Second count on sample

$$y = \frac{y_1 + y_2}{2}$$

This test can be applied to recounts from the same laboratory that produced the data in Table 13.1. Let y_1 = 346 fibers per square millimeter and y_2 = 382 fibers per square millimeter. Then:

$$y_1 = \sqrt{346} = 18.6$$

$$y_2 = \sqrt{382} = 19.5$$

$$\bar{y} = \frac{18.6 + 19.5}{2} \quad 19.0$$

and

$$|18.6 - 19.5| = 0.9$$

$$(2.8)\,(\bar{y})\,(CV) = (2.8)\,(19.0)\,(0.10) = 5.32$$

Since 0.9 is less than 5.3, the two counts pass the test, and the difference between the two is considered to be acceptable.

Another test that can be used in a quality control program, when there are no specific procedures designated, is the Youden Ranking Test for laboratories (Youden and Steiner). This test was previously discussed in Chapter 11. Its use has been proposed as an approach to dealing with the OSHA asbestos standard requirement for an inter-laboratory quality control program (Tombes and Calpin). An example of the ranking test for laboratories follows. In Table 13.2, data for three laboratories and nine samples are presented. These data are actual asbestos counts provided by three different laboratories for the same samples. These data represent total fiber counts per filter.

From Table 11.7 (Chapter 11), find that the allowable scores for a combination of four laboratories and seven samples are 10 and 25. Since none of the four laboratories has a score that is lower than 10 or higher than 25, the conclusion is that there is no systematic error present that significantly affects the sample results. If one of the laboratories did have a score lower than 10 or greater than 25, the conclusion would be that there was some sort of systematic bias that affected the asbestos results for that laboratory. It would then be the responsibility of the laboratory director or other supervisor to determine the reasons for the anomaly in the laboratory's performance.

Table 11.8 is structured so that the likelihood that a laboratory will have a score that is either less than or equal to the lower tail limit or greater than or equal to the upper tail limit, due strictly to chance alone, is 5%. When a laboratory's score falls

Table 13.2 Youden Ranking Test for Laboratories

| Lab ID | Number of Samples | | | | | | | Rank Sum |
	1	2	3	4	5	6	7	
D	106000	129000	245000	149000	205000	144000	158000	
D Rank	3	2	2	4	3	1	4	19
E	76000	126000	158000	143000	104000	153000	14000	
E Rank	1	1	1	3	1	3	2	12
F	87000	206000	277000	130000	216000	144000	141000	
F Rank	2	4	3	2	4	2	3	20
G	117000	169000	306000	107000	181000	169000	132000	
G Rank	4	3	4	1	2	4	1	19

outside either of these limits, the assumption is made that the result is due to some systematic error, rather than to chance alone.

The procedure for conducting this test is quite simple and straightforward.

1. Distribute at least four samples to three or more laboratories. The samples should all be analyzed for the same analyte.
2. List the results for each laboratory and sample, in a convenient manner, such as in Table 13.2.
3. Rank all the laboratories for each sample. The laboratory that has the lowest value for a sample is assigned a ranking of 1, and the laboratory that has the has the highest value is assigned a ranking of n, where n represents the number of laboratories. Ties are broken randomly (e.g., flipping a coin).
4. For each laboratory, sum the rank values for all samples.
5. Select the appropriate upper and lower limits from Table 11.7, using the number of laboratories and the number of samples analyzed in the study. The example above has four laboratories and seven samples.
6. Compare the rank sum for each laboratory to the upper and lower limits found in Step 5. If the rank sum is less than or equal to the lower limit, or greater than or equal to the upper limit, it must be concluded that there is a systematic bias in the analytical system for that laboratory.

This test does not necessarily involve any objective reference material, although it may. It says nothing about the ability of the laboratory to produce accurate results. Successful performance indicates only that a laboratory performs in a manner consistent with other laboratories in the analysis of a specific set of samples.

REFERENCES

Abell, M. T., S. Shulman, and D. Baron. 1989. The quality of fiber count data. Applied Industrial Hygiene 4:273.

Tombes, C., and J. A. Calpin. 1989. A simple quality control system for evaluation of inter-laboratory differences in fiber counting in accordance with NIOSH 7400 method. American Industrial Hygiene Association Journal 49:A695.

Youden, W. J., and E. H. Steiner. 1987. Statistical Manual of the Association of Official Analytical Chemists. Arlington, VA: Association of Official Analytical Chemists.

14

Nonroutine Quality Control Procedures

Traditionally, quality control procedures have been directed at situations that can be characterized as routine analytical work—the kind of work that involves analyzing or testing large numbers of identical samples, on a regular basis. In-house laboratories and clinical laboratories are facilities where this kind of routine analytical or testing work is likely to be performed. Routine work can be contrasted with situations, where a large portion of the laboratory's work is comprised of small numbers of a large variety of sample types. The control of quality in this instance becomes a problem because the cost of developing an effective quality procedure for a small number of samples cannot be amortized over a long period of time and a large number of samples.

Nevertheless, the need for some sort of evaluation of sample results is a must in this case. It is not always true that earthshaking decisions rest on every set of sample values reported by a laboratory, but, even if the issue has only a minimal economic effect, this can be significant in a small business. Numerous decisions are made daily on the basis of analytical results produced by all kinds of laboratories. For instance, small businesses may make decisions to add pollution control equipment as a result of reported analytical data. Or, on the other hand, they may make a decision not to add pollution control equipment or protective equipment for workers when they should, because of the data reported by an analytical laboratory. These kinds of decisions can be costly, not only because of the capital cost of the equipment, if the decision is "to buy," but also because of the legal cost if the decision is to "not buy," and it is the wrong decision.

The cost of quality control becomes critical with these kinds of samples. If the laboratory does not have extensive experience with a procedure, the customer might be paying for some method development work, as well as for the quality control work, if any is done. This means that either the unit cost per analysis will be extraordinarily high to the customer, or the laboratory will have to absorb substantial costs and hope that it can recover them in subsequent business with the customer.

Some of the elements of small sample number quality control that can be used by analytical laboratories are sample spikes, replicate analyses, reference samples, and blank determinations. These are things that can be done by laboratory personnel, to evaluate their own performance. These are also things that might be done by the customer, to investigate the performance of the laboratory.

SPIKING

Spiking samples, to get an idea of the quality of the results produced by a laboratory, is a technique that has been used for a long time. Spiking samples is also known as the method of additions. This procedure is carried out by adding known quantities of the analyte to an actual sample. Usually four aliquots of a sample are used—one has nothing added to it; the second, third, and fourth have known, increasing quantities of the analyte added to them. The analyte (usually in a standard solution) are added to the aliquots, in quantities that approximate the quantity of analyte thought to be present in the original sample.

The four samples are then analyzed and the results plotted against instrument response. The aliquots with the added material will, of course, plot as the original concentration, plus the added material. The unadulterated sample will be plotted as 0 on the graph (Figure 14.1). A straight line is then drawn through the points, to its intersection with the x-axis. The original concentration of material can be determined by measuring the distance along the x-axis, between the point of intersection and the origin. This procedure allows the analyst to determine the original concentration of material in the particular sample chosen for spiking, and to evaluate the ability of the method to successfully recover the desired analyte from the sample. There are a number of factors that can affect the successful implementation of the procedure. If the original sample matrix affects the availability of the analyte, the spiking procedure may not detect this, resulting in an erroneous analysis. Inaccuracies in preparing the spiked samples can seriously affect the slope of the calibration line, causing errors to be made in determining the original concentration.

If a straight line cannot be plotted through the data points, there is obviously something wrong with some element of the procedure. If the recoveries of the added analyte do not appear to be acceptable, there must be some doubt about the ability of the method to adequately recover the original sample analyte, for the original sample matrix. The failure of this approach for a given method of analysis may be indicative of a need to find a different procedure for the samples in question.

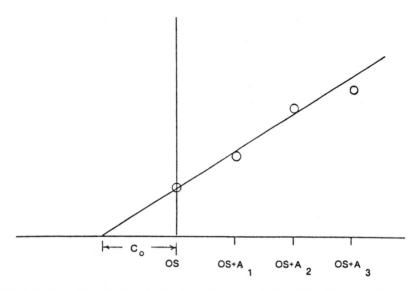

Method of Additions. OS = original sample. OS + A_1 = original sample plus first addition. OS + A_2 = original sample plus second addition. C_o = Graphically determined concentration of original sample.

REPLICATE ANALYSIS

One traditional method of quick and dirty quality control has been to divide a sample into two aliquots and to compare the results for the two aliquots. This comparison is usually intuitive. The analyst decides beforehand what an acceptable difference will be, and evaluates the two sample results, on that basis. This type of evaluation must be done with great caution, however. It cannot be used as a sole determinate of acceptability. In other words, the ability to duplicate a result does not guarantee adequacy of the method or the analyst, because a bad result can be duplicated just as easily as a good one.

How then, can one approach the problem of duplicate evaluation when there is a paucity of data, and the laboratory has no history for the particular method chosen for the analysis? This is a question that is not readily answered. One approach is to use the method of Abell et al., which was originally developed for use with asbestos samples. However, there is no reason why it should not be applicable to any sample analysis.

This simple test requires that the user have some knowledge of the variance of the procedure. This knowledge can be obtained from some outside source, such as an analytical methods manual, which reports this data for various methods. Of course, this presumes that the analyst is going to use the method that has a reported variance. It also presumes that the analyst will follow the method as closely as possible. A methods manual that reports this kind of information is the National Institute for Occupational Safety and Health (NIOSH) Manual of Analytical Methods.[1] This manual presents a coefficient of variation (CV) for many of the methods that it contains.

The test proposed by Abell et al. is:

$$|y_1 - y_2| > 2.77 \, (\bar{y}) \, CV \qquad (14\text{-}1)$$

where:

$$y_1 = \text{first result}$$
$$y_2 = \text{second result}$$

CV is the historical coefficient of variation

If the difference between the two values exceeds the calculated value of the right side of the expression, it is probable that one of the counts is biased, and the analyst should begin an investigation, to determine the reasons for the difference. It can be seen that this test is specifically designed to determine if the sample variance is significantly different from the know population variance. The derivation of this expression can be found in the Abell et. al. paper.

This test is nothing more than a way to get a handle on the performance of a given procedure. It is not designed to provide an absolute evaluation of the quality of a particular method/analyst combination. It is also not designed to take the place of control charts or other statistical tests, but merely to supplement them in short-term situations. A weakness of this test is the forced reliance on a published coefficient of variation, rather than one developed by the laboratory, using the method. Its decision-making power increases, as knowledge of the actual coefficient of variation increases.

Two additional caveats are in order:

1. Errors associated with the production of the aliquot or aliquots may be a problem.
2. Because the test is based only on replicates, it does not have much power. More power may be achieved if multiple aliquots and a more generalized test are used.

REFERENCE SAMPLES

Reference sample usage for short sample runs is an invaluable way to provide a quality control tool for evaluating the data. Morroti (1990) has determined that using reference samples markedly improves the performance of analytical laboratories that deal with hazardous waste samples. While this procedure is certainly a desirable one, it is not always achievable. Reference samples pertaining to a particular method under investigation may not be available or, if available, may be unreasonably expensive. If at all possible, every sample set should be accompanied by one or more analyses of an accepted reference sample.

BLANK DETERMINATIONS

Operating a successful quality control program is contingent upon a number of factors, one of which is an understanding of and an appreciation of the necessity for incorporating blank analyses into the program. A blank is a sample that is prepared by the laboratory or an outside supplier and that contains no analyte. If the blank contains no analyte, there should be no instrumental response when the blank is subjected to the analysis. If there is an instrumental response, the analyst must decide how to deal with it. The most common approach is to subtract the blank value from the sample values or to use the blank to set the low threshold for instrument response.

If the blank value is too large, it becomes a problem that cannot be handled by simple subtraction, since one must assume that the matrix effect or contamination is consistent for all the samples being reported. The analyst must decide how large a blank value is unacceptable. In some cases, there may be some guidance given in the method itself. For example, the OSHA regulation 29CFR1910.1001 mandates a specific analytical procedure for analyzing of air samples for asbestos fibers. In the discussion of the procedural steps for the method, an allowable blank fiber count is specified. If the blank count is greater than that allowed, the samples that are represented by the blank may be unacceptable.

A feeling for acceptable blank values can be gained by control charting the blank values. Taylor (1987) gives a good description of how to handle sample values that require a blank correction. This approach is a conservative one that results in a sample value that is actually higher than it should be. If the sample is one that is part of a regulatory compliance study, the analyst may be unfairly penalizing the client.

Blanks can be of two types: reagent blanks, which are used to evaluate the analytical procedure itself, including the suitability of the reagents, the manipulative steps used by the analyst, the competence of the analyst, and the cleanliness of the equipment; or field blanks, which are designed to test the handling, storage, and sampling collection procedures that are undergone by the samples. The analyst usually has direct input only to the reagent blank.

BLIND SAMPLES

Blind samples are samples whose identity is known, but whose value is known only to the individual submitting them. Thus, the analyst knows that he or she is working on a control sample, but does not know what the quantity of analyte in the sample is. Blind samples can be useful, but they generally require that a second person participate in the analytical process as a quality control coordinator.

OTHER ANALYTICAL TECHNIQUES

One approach that is accepted as a quality control tool is to use a second analytical procedure, to confirm a result produced by another procedure or method. If the two results are in close agreement, it is assumed that the value reported is accurate. The two methods used must be different in principle. For example, a colorimetric analytical method for lead could be used to confirm an atomic absorption result.

REFERENCES

Abell, M. T., S. Shulman, and P. Baron. 1989. The quality of fiber count data. Applied Industrial Hygiene 4:273.

Morotti, J. M. 1990. Quality control: determining real-world performance. Environmental Laboratories 2:26.

Taylor, John K. 1987. Quality Assurance of Chemical Measurements. Chelsea, MI: Lewis Publishers, Inc.

Appendix

Table of Critical Values for $S_{2n-1,n}/S^2$ or $S^2_{1,2}/S^2$ for Simultaneously Testing the Two Largest or Two Smallest Observations

No. of Obs. n	Lower .1% Sig. Level	Lower .5% Sig. Level	Lower 1% Sig. Level	Lower 2.5% Sig. Level	Lower 5% Sig. Level	Lower 10% Sig Level
4	.0000	.0000	.0000	.0002	.0008	.0031
5	.0003	.0018	.0035	.0090	.0183	.0376
6	.0039	.0116	.0186	.0349	.0564	.0920
7	.0135	.0308	.0440	.0708	.1020	.1479
8	.0290	.0563	.0750	.1101	.1478	.1994
9	.0489	.0851	.1082	.1492	.1909	.2454
10	.0714	.1150	.1414	.1864	.2305	.2863
11	.0953	.1448	.1736	.2213	.2667	.3227
12	.1198	.1738	.2043	.2537	.2996	.3552
13	.1441	.2016	.2333	.2836	.3295	.3845
14	.1680	.2280	.2605	.3112	.3568	.4106
15	.1912	.2530.	.2859	.3367	.3818	.4345
16	.2136	.2767	.3098	.3603	.4048	.4562
17	.2350	.2990	.3321	.3822	.4259	.4761
18	.2556	.3200	.3530	.4025	.4455	.4944
19	.2752	.3398	.3725	.4214	.4636	.5113
20	.2939	.3585	.3909	.4391	.4804	.5270
21	.3118	.3761	.4082	.4556	.4961	.5415
22	.3288	.3927	.4245	.4711	.5107	.5550
23	.3450	.4085	.4398	.4857	.5244	.5677
24	.3605	.4234	.4543	.4994	.5373	.5795
25	.3752	.4376	.4680	.5123	.5495	.5906

$$S^2 = \sum_{i=1}^{n} (x_i - X)^2; \; \overline{x} = \frac{1}{n} \sum_{i=1}^{n} X_i \; ; x_1 \leq x_2 \leq \ldots \leq x_n$$

$$S_{1,2}^2 = \sum_{i=2}^{n} (x_i - \overline{x}_{1,2})^2; \; \overline{x}_{1,2} = \frac{1}{n-2} \sum_{i=2}^{n} X_i$$

$$S_{n-1,n}^2 = \sum_{i=1}^{n-2} (x_i - \overline{x}_{n-1,n})^2 : \overline{x}_{n-1,n} = \frac{1}{n-2} \sum_{i=1}^{n-2} x_i$$

Table of Critical Values for $S_{2n-1,n}/S^2$ or $S^2_{1,2}/S^2$ for Simultaneously Testing the Two Largest or Two Smallest Observations

No. of Obs. n	Lower .1% Sig. Level	Lower .5% Sig. Level	Lower 1% Sig. Level	Lower 2.5% Sig. Level	Lower 5% Sig. Level	Lower 10% Sig Level
26	.3893	.4510	.4810	.5245	.5609	.6011
27	.4027	.4638	.4933	.5630	.5717	.6110
28	.4156	.4759	.5050	.5470	.5819	.6203
29	.4279	.4875	.5162	.5574	.5916	.6292
30	.4397	.4985	.5268	.5672	.6008	.6375
31	.4510	.5091	.5369	.5766	.6095	.6455
32	.4618	.5192	.5465	.5856	.6178	.6530
33	.4722	.5288	.5557	.5941	.6257	.6602
34	.4821	.5381	.5646	.6023	.6333	.6671
35	.4917	.5469	.5730	.6101	.6405	.6737
36	.5009	.5554	.5811	.6175	.6474	.6800
37	.5098	.5636	.5889	.6247	.6541	.6860
38	.5184	.5714	.5963	.6316	.6604	.6917
39	.5266	.5789	.6035	.6382	.6665	.6972
40	.5345	.5862	.6104	.6445	.6724	.7025
41	.5422	.5932	.6170	.6506	.6780	.7076
42	.5496	.5999	.6234	.6565	.6834	.7125
43	.5568	.6064	.6296	.6621	.6886	.7172
44	.5637	.6127	.6355	.6676	.6936	.7218
45	.5704	.6188	.6412	.6728	.6985	.7261
46	.5768	.6246	.6468	.6779	.7032	.7304
47	.5831	.6303	.6521	.6828	.7077	.7345
48	.5892	.6358	.6573	.6876	.7120	.7384
49	.5951	.6411	.6623	.6921	.7163	.7422
50	.6008	.6462	.6672	.6966	.7203	.7459
51	.6063	.6512	.6719	.7009	.7243	.7495
52	.6117	.6560	.6765	.7051	.7281	.7529

Table of Critical Values for $S_{2n-1,n}/S^2$ or $S^2_{1,2}/S^2$ for Simultaneously Testing the Two Largest or Two Smallest Observations

No. of Obs. n	Lower .1% Sig. Level	Lower .5% Sig. Level	Lower 1% Sig. Level	Lower 2.5% Sig. Level	Lower 5% Sig. Level	Lower 10% Sig Level
53	.6169	.6607	.6809	.7091	.7319	.7563
54	.6220	.6653	.6852	.7130	.7355	.7595
55	.6269	.6697	.6894	.7168	.7390	.7627
56	.6317	.6740	.6934	.7205	.7424	.7658
57	.6364	.6782	.6974	.7241	.7456	.7687
58	.6410	.6823	.7012	.7276	.7489	.7716
59	.6454	.6862	.7049	.7310	.7520	.7744
60	.6497	.6901	.7086	.7343	.7550	.7772
61	.6539	.6938	.7121	.7375	.7580	.7798
62	.6580	.6975	.7155	.7406	.7608	.7824
63	.6620	.7010	.7189	.7437	.7636	.7850
64	.6658	.7045	.7221	.7467	.7664	.7874
65	.6696	.7079	.7253	.7496	.7690	.7898
66	.6733	.7112	.7284	.7524	.7716	.7921
67	.6770	.7144	.7314	.7551	.7741	.7944
68	.6805	.7175	.7344	.7578	.7766	.7966
69	.6839	.7206	.7373	.7604	.7790	.7988
70	.6873	.7236	.7401	.7630	.7813	.8009
71	.6906	.7265	.7429	.7655	.7836	.8030
72	.6938	.7294	.7455	.7679	.7859	.8050
73	.6970	.7322	.7482	.7703	.7881	.8070
74	.7000	.7349	.7507	.7727	.7902	.8089
75	.7031	.7376	.7532	.7749	.7923	.8108
76	.7060	.7402	.7557	.7772	.7944	.8127
77	.7089	.7427	.7581	.7794	.7964	.8145
78	.7117	.7453	.7605	.7815	.7983	.8162
79	.7145	.7477	.7628	.7836	.8002	.8180
80	.7172	.7501	.7650	.7856	.8021	.8197
81	.7199	.7525	.7672	.7876	.8040	.8213
82	.7225	.7548	.7694	.7896	.8058	.8230
83	.7250	.7570	.7715	.7915	.8075	.8245
84	.7275	.7592	.7736	.7934	.8093	.8261
84	.7300	.7614	.7756	.7953	.8109	.8276
86	.7324	.7635	.7776	.7971	.8126	.8291
87	.7348	.7656	.7796	.7989	.8142	.8306
88	.7371	.7677	.7815	.8006	.8158	.8321
89	.7394	.7697	.7834	.8023	.8174	.8335
90	.7416	.7717	.7853	.8040	.8190	.8349
91	.7438	.7736	.7871	.8057	.8205	.8362
92	.7459	.7755	.7889	.8073	.8220	.8376
93	.7481	.7774	.7906	.8089	.8234	.8389
94	.7501	.7792	.7923	.8104	.8248	.8402

Table of Critical Values for $S_{2n-1,n}/S^2$ or $S^2{}_{1,2}/S^2$ for Simultaneously Testing the Two Largest or Two Smallest Observations

No. of Obs. n	Lower .1% Sig. Level	Lower .5% Sig. Level	Lower 1% Sig. Level	Lower 2.5% Sig. Level	Lower 5% Sig. Level	Lower 10% Sig Level
95	.7522	.7810	.7940	.8120	.8263	.8414
96	.7542	.7828	.7957	.8135	.8276	.8427
97	.7562	.7845	.7973	.8149	.8290	.8439
98	.7581	.7862	.7989	.8164	.8303	.8451
99	.7600	.7879	.8005	.8178	.8316	.8463
100	.7619	.7896	.8020	.8192	.8329	.8475
101	.7637	.7912	.8036	.8206	.8342	.8486
102	.7655	.7928	.8051	.8220	.8354	.8497
103	.7673	.7944	.8065	.8233	.8367	.8508
104	.7691	.7959	.8080	.8246	.8379	.8519
105	.7708	.7974	.8094	.8259	.8391	.8530
106	.7725	.7989	.8108	.8272	.8402	.8541
107	.7742	.8004	.8122	.8284	.8414	.8551
108	.7758	.8018	.8136	.8297	.8425	.8563
109	.7774	.8033	.8149	.8309	.8436	.8571
110	.7790	.8047	.8162	.8321	.8447	.8581
111	.7806	.8061	.8175	.8333	.8458	.8591
112	.7821	.8074	.8188	.8344	.8469	.8600
113	.7837	.8088	.8200	.8356	.8479	.8610
114	.7852	.8101	.8213	.8367	.8489	.8619
115	.7866	.8114	.8225	.8378	.8500	.8628
116	.7881	.8127	.8237	.8389	.8510	.8637
117	.7895	.8139	.8249	.8400	.8519	.8646
118	.7909	.8152	.8261	.8410	.8529	.8655
119	.7923	.8614	.8272	.8421	.8539	.8664
120	.7937	.8176	.8284	.8431	.8548	.8672
121	.7951	.8188	.8295	.8441	.8557	.8681
122	.7964	.8200	.8606	.8451	.8567	.8689
123	.7977	.8211	.8317	.8461	.8576	.8697
124	.7990	.8223	.8327	.8471	.8585	.8705
125	.8003	.8234	.8338	.8480	.8593	.8715
126	.8016	.8245	.8348	.8490	.8602	.8721
127	.8028	.8256	.8359	.8499	.8611	.8729
128	.8041	.8267	.8369	.8508	.8619	.8737
129	.8053	.8278	.8379	.8517	.8627	.8744
130	.8065	.8288	.8389	.8526	.8636	.8752
131	.8077	.8299	.8398	.8535	.8644	.8759
132	.8088	.8309	.8408	.8544	.8652	.8766
133	.8100	.8319	.8718	.8553	.8660	.8773
134	.8111	.8329	.8427	.8561	.8668	.8780
135	.8122	.8339	.8436	.8570	.8675	.8787
136	.8134	.8349	.8445	.8578	.8683	.8794

Table of Critical Values for $S_{2n-1,n}/S^2$ or $S^2_{1,2}/S^2$ for Simultaneously Testing the Two Largest or Two Smallest Observations

No. of Obs. n	Lower .1% Sig. Level	Lower .5% Sig. Level	Lower 1% Sig. Level	Lower 2.5% Sig. Level	Lower 5% Sig. Level	Lower 10% Sig Level
137	.8145	.8358	.8454	.8586	.8690	.8801
138	.8155	.8368	.8463	.8594	.8698	.8808
139	.8166	.8377	.8472	.8602	.8705	.8814
140	.8176	.8387	.8481	.8610	.8712	.8821
141	.8187	.8396	.8489	.8618	.8720	.8827
142	.8197	.8405	.8498	.8625	.8727	.8834
143	.8207	.8414	.8506	.8633	.8734	.8840
144	.8218	.8423	.8515	.8641	.8741	.8846
145	.8227	.8431	.8523	.8648	.8747	.8853
146	.8237	.8440	.8531	.8655	.8754	.8859
147	.8247	.8449	.8539	.8663	.8761	.8865
148	.8256	.8457	.8547	.8670	.8767	.8871
149	.8266	.8465	.8555	.8677	.8774	.8877

An observed ratio less than the appropiate critical ratio in this table calls for rejection of the null hypothesis.

The values are taken from Grubbs and Beck 1972. This table is taken from Grubbs 1979 pages 3-42-45. By permission. U.S. Army Statistics Manual. DARCOM, p. 706-1-3

Index

Acceptable quality level, sampling
plans, 70-71, 75-83
Analysis of variance, 66
ANSI/ASQC Standard Z1.4-1981, 69,
70, 75, 76, 80
See also Sampling plans
Area of indecision, sampling plans, 80
Asbestos counting. *See* Statistical
quality control
Average outgoing quality, sampling
plans, 73-75

Bathtub distribution, 22, 24
Between group variability, 29, 30, 33
Bias, meaning of, 29-30
Blank determinations, 164
types of blanks, 164
Blind samples, 164-165

Central limit theorem, 20
and population distributions, 100
Chi-square test
difference between sample variability and population variability,
56-58

as normality test, 27
outlier test based on, 102
Coefficient of variation, 16
statistical quality control, 154-157,
163
Consumer's risk sampling plans, 85
Continuous data set outliers
control charts, 143-144
duplicate measurements, 142-143
runs, use of, 144-147
triplicate measurements, 143
Control charts
basic assumptions related to, 30-33
calculation of control limits, 35-36
calculation on warning limits, 36-37
calculations, 35
continuous data set outliers, 143-144
control limits, determination of, 31-
32, 34
cumulative sum control charts, 44-47
example, chart for reagent blank
determinations, 37-40
fraction defective charts, 47-48
for individual values, 41
moving averages, 42-43
moving ranges, 43

out-of-control points, 143
procedures for construction of,
    33–35
Shewhart Control Chart, 28–29
sigma charts, 44
uses of, 29
variable subgroup sizes, 44
Correlation
    Correlation Coefficient, 92–93
    and slope of line, 86
    Standard Error of the Estimate, 90–92
Correlation Coefficient, calculation
    of, 92– 93
Cumulative Normal Distribution, out-
    lier test based on, 100, 102
Cumulative sum control charts,
    44–47

Degrees of freedom, 15
Determinate errors, 29
Distribution curve
    bathtub distribution, 22, 24
    kurtosis, 8
    lognormal distribution, 21
    measure of central tendency, 8
    measure of variability, 8
    Poisson distribution, 22
    skewed curve, 8
    unimodal and multimodal distri-
        butions, 8
Distribution free tests, and population
    distributions, 100
Dodge-Rohmig Sampling Inspection
    Tables, 85

Errors, determinate and indeterminate,
    29
Extreme rank sum test, multiple outliers,
    137–139

Field blanks, 164
Fraction defective charts, 47–48
Frequency distributions, 6–9
    characteristics of distribution
        curve, 6–8
    Gaussian distribution, 6
    variations in distribution curve, 9

Gap and straggler test, 66
Gaussian distributions, 6
    See also Normal distribution

Histograms
    distribution histogram, 7
    as normality test, 23, 27

Indeterminate errors, 29

Kurtosis, distribution curve, 8
Kurtosis test, multiple outliers, 136

Least Squares Regression Line, calcu-
    lation of, 86–90
Linear regression, 86, 90
    Least Squares Regression Line, 86–
        90
Lognormal distribution, 21

Mean, 9–10
    trimmed mean, 129
Measures of central tendency, 8
    mean, 9–10
    median, 10
    mode, 10–11
Measures of variability, 8, 11–12

need to measure variability, 13-14
standard deviation, 14-17
variance, 11
Median, 10
Method of additions, 161
MIL-STD-105D, 69, 70, 75-83
See also Sampling plans
Mode, 10-11
Moving averages, 42-43
Moving ranges, 43
Multimodal distributions, 8
Multiple outliers
  kurtosis test, 136
  rejection of multiple outliers, 129-135
  skewness test, 136
  slippage tests, 136-137
  test for one high and one low outlier, 125-126
  test for two high or two low outliers, 126-129
  Youden extreme rank sum test, 137-139

Normal distributions
  central limit theorem, 20
  variations in, 17-19
Normality tests
  chi-square test, 27
  histograms, 23, 27
Normalizing transformations, and population distributions, 100

Operating characteristic curve, sampling plans, 70
Outliers
  base population characteristics, 98-100
  choice of outlier rejection test, 100
  purposes of detecting outliers, 95-98

reasons for rejection of, 94
test based on chi-square, 102
test based on Cumulative Normal Distribution, 100, 102
See also Continuous data set outliers; Multiple outliers; Single outliers
Outlying observations. *See* Outliers
Out-of-control points, 143

Poisson distribution, 22
Polygon, distribution polygon, 7
Population contamination, 105
Population distributions
  and central limit theorem, 100
  and distribution free tests, 100
  and normalizing transformations, 100
Probabilities, 1-6
  random number table, guidelines for use, 4-6
Producer's risk schemes, 69

Quality control
  blank determinations, 164
  blind samples, 164-165
  confirming results with second procedure, 165
  cost factors, 160-161
  reference samples, 163-164
  replicate analysis, 162-163
  and routine analytical work, 160
  spiking, 161
  subjective quality control, 154
  See also Statistical quality control

Randomness, tests for, 95
Random number table, guidelines for use, 4-6

Range, to estimate standard deviation, 65
Reagent blanks, 164
Reference samples, 163-164
Regression
  Least Squares Regression Line,
    calculation of, 86-90
  linear regression, 86, 90
Relative standard deviation, 16
Replicate analysis, 162-163
Rosner sequential test, 131
Runs
  approaches to study of, 144-146
  definition of, 144
  use of, 144-147

Sampling plans, 69-85
  acceptable quality level, 70-71, 75-
    83
  ANSI/ASQC Standard Z1.4-1981,
    69, 70, 75, 76, 80
  area of indecision, 80
  average outgoing quality, 73-75
  classification of defects, 76-77
  consumer's risk sampling plans, 85
  Dodge-Rohmig Sampling Inspec-
    tion Tables, 85
  MIL-STD-105D, 69, 70, 75-83
  most commonly used plans, 69
  operating characteristic curve, 70
  producer's risk schemes, 69
  reasons for use of, 69
  for sequential sampling, 83
  taking fixed sample size, 75
Shewhart Control Chart, 28-29
Sigma charts, 44
Single outliers
  population contamination, 105
  situations related to, 104
  test when μ and σ are unknown,
    119-124

test when μ and σ are unknown but
    have estimate of σ_2, 111-119
test when only * is known, 105-
    111
Skewed curve, positively and nega-
    tively skewed, 8
Skewness test, multiple outliers, 136
Slippage tests, 136-137
  k-sample slippage test, 136-137
Slope, and correlation, 86
Spiking, 161
Standard deviation
  population standard deviation, 14-
    15
  relative standard deviation, 16
  sample standard deviation, 15
  and use of programmable calcula-
    tors, 16-17
Standard Error of the Estimate, calcu-
    lation of, 90-92
Statistical quality control
  categories of procedures, 152
  coefficient of variation, 154-157,
    163
  legislated requirements, 153-154
  quality control program, 153
  requirements for, 151
  traditional techniques, 152-153
  Youden Ranking Test, 157-159

Tests for significance of differences,
    49-68
  chi-square test, difference between
    sample variability and popula-
    tion variability, 56-58
  "F" test for difference in variability
    in two samples, 58
  between sample mean and popula-
    tion mean, 50-53
  between two sample means, 55

between two sample means based on independent random samples, 54

Unimodal distributions, 8

Variability. *See* Measures of variability
Variance, 11
    trimmed variance, 129

Warning limits, control charts, 32, 36–37
Within group variability, 29, 30, 33

Youden extreme rank sum test, multiple outliers, 137–139
Youden Ranking Test, 157–159